ADVANCES IN ENVIRONMENTAL RESEARCH

VOLUME 67

ADVANCES IN ENVIRONMENTAL RESEARCH

Additional books and e-books in this series can be found on Nova's website under the Series tab.

ADVANCES IN ENVIRONMENTAL RESEARCH

ADVANCES IN ENVIRONMENTAL RESEARCH

VOLUME 67

JUSTIN A. DANIELS
EDITOR

NOTICE TO THE READER

Library of Congress Cataloging-in-Publication Data

ISBN: 978-1-53615-009-4
ISSN: 2158-5717

Published by Nova Science Publishers, Inc. † New York

CONTENTS

PREFACE

This book highlights current knowledge on iodide mobility in a soil-plant system and provides a theoretical and experimental basis for a better understanding of the geochemical behaviour of iodine in soils, including its availability in the food chain.

Following this, the authors provide a comprehensive overview of the uses of iodide salts by focusing on their applications in plants and microbiology, mechanisms of action and possible new uses in histology.

The penultimate study contributes to a better understanding of Piauí's vegetation through a floristic survey, phytosociological study and knowledge about the economic potential and geographic distribution of species from an area belonging to the municipality of Brasileira, north of Piauí, Brazil. Fifty five families, 126 genera and 141 species represented the flora.

Biosphere reserves face the challenge of sustainable development. They have to foster economic development that is ecologically and culturally sustainable. Paradoxically, the demographic-economic-entrepreneurial nexus of biosphere reserves has not been researched, an omission addressed in the closing chapter by studying the towns of the Gouritz Cluster Biosphere Reserve in South Africa.

Chapter 1 - Iodine is an essential micronutrient for humans because it is necessary for the synthesis of thyroid hormones. The most abundant iodine species are iodide and iodate, from which iodide represents the most bioavailable form for hormone production. Despite the importance of iodine, only a few studies to date have dealt with its distribution and mobility in the environment. The primary source of iodine is soil. From soil, it is translocated into plants, where the upper levels of the food chain are enriched. The translocation of iodine from soil depends on various (bio)geochemical factors, that influence iodine's mobility by affecting the bioavailable fraction for primary producers. The main criteria for iodide availability are water solubility, sorption to soil components and microbial activity. The presence of soil microorganisms prevents or limits the plant's uptake of iodine by its microbial accumulation. Furthermore, microorganisms can metabolically transform bioaccumulated iodine into volatile derivatives, such as methyl-iodide, and, thus, alter the content of bioavailable iodide in soils. There are also several other biogeochemical factors in soil that contribute to the change of iodine speciation in soil. Some soil redox active constituents can efficiently transform iodate into iodide or vice versa. The same affect can be ascribed to the metabolic activity of plant roots or microorganisms. Therefore, the authors present in this chapter a review on the geochemical behaviour of iodide (and iodate) in soils. Most of the information provided is accompanied with the authors' own experimental data on soil-plant-microbial interactions with iodine, which provide new insight into iodine mobility in soils, as well as its bioavailability to plants and filamentous fungi. Experimentally, bioaccumulation efficiency of both readily bioavailable water soluble iodine species and a stabilised (aged) iodine from soil, which had previously gone through various geochemical transformations including sorption onto soil particles, were evaluated. The accumulated amount of iodine, as well as its distribution in soils, was determined by various one-step or sequential extraction procedures, as well as a few other methods of analytical geochemistry. The authors also provide new information on factors affecting soil-to-plant transfer factor and iodine phytotoxicity. Biologically induced loss of iodine via volatilisation into the atmosphere, which is caused by the biochemical transformation of iodine into volatile species by plants and microscopic filamentous fungi, are also evaluated.

This chapter therefore, highlights current knowledge on iodide mobility in a soil-plant system and provides a theoretical and experimental basis for better understanding of the geochemical behaviour of iodine in soils, including its availability in the food chain.

Chapter 2 - The study of the microscopic structure of the biological material, which allows us to know how the unitary components are structurally and functionally related, has always been of paramount interest in the biomedical research. That knowledge is at the intersections between biochemistry, molecular biology and physiology on the one hand and pathological processes and their consequences on the other. Since the earliest years of the XIX century, several salts of iodide such as potassium iodide (Lugol) or more recently the Propidium iodide (PI) have been used in animal, plant, bacterial and fungal histology. Their application range goes from the detection of simple molecules to determine their content in a given cell or tissue (e.g., detection of polysaccharides in pollen of *Pinus pinea*), to the determination of physiological statuses (e.g., fungal spore viability of *Rhizopogon roseolus*). This work intends to provide a comprehensive overview of the uses of the iodide salts by especially focussing on its application in plants and microbiology, the mechanisms of action and the possible new uses in histology.

Chapter 3 - This study aimed to contribute to a better knowledge of Piauí´s vegetation through a floristic survey, phytosociological study and knowledge about the economic potential and geographic distribution of species from an area belonging to the municipality of Brasileira, north of Piauí, Brazil. Fifty five families, 126 genera and 141 species represented the flora. The families with the highest species richness were Fabaceae, Lamiaceae, Malvaceae, Apocynaceae and Bignoniaceae. In the study area, 35 species, according to "Flora of Brazil 2020," were registered as endemic to Brazil. The species that had the highest importance values (VI) in the studied plant community were *Ephedranthus pisocarpus*, *Copaifera langsdorffii*, *Myrcia guianensis*, *Terminalia fagifolia*, *Parkia platycephala*, *Astrocaryum campestre* and *Anacardium occidentale*. The Shannon Index (H ') was 2.85 nats ind.-1 and the Equability Index (J') was 0.696. The maximum, mean and minimum heights of the individuals were 15m, 7.6m and 3.4 m and the maximum, average and minimum diameters were 47.7 cm, 9.2 cm and 5 cm, respectively. Fifty seven species distributed in 32 families present economic potential, being the majority of medicinal and meliferous use, being *Byrsonima correifolia*, *Caryocar coriaceum*, *Handroanthus impetiginosus*, *Magonia pubescens* and *Parkia platycephala* the ones with greater number of use in the literature. *Combretum leprosum*, *Cereus jamacaru* and *Bauhinia ungulata* were widely distributed in all plant formations compared to the local flora.

Chapter 4 - Biosphere reserves face the challenge of sustainable development. They have to foster economic development that is ecologically and culturally sustainable. Paradoxically, the demographic-economic-entrepreneurial nexus of biosphere reserves has not been researched, an omission addressed here by studying the towns of the Gouritz Cluster Biosphere Reserve in South Africa. There is extensive orderliness in the above nexus and the interlinkages of many of its characteristics have been quantified. Systematic regularities, some time-independent, occur in the socioeconomic domain and between population sizes of towns and their enterprise numbers. Power-laws describe population-based scaling of a number of characteristics, indicating differences between small and large towns. Enterprise profiles have changed over time, with the tourism and hospitality services sector the biggest winner and the trade services sector the biggest loser. Productive knowledge, measured as enterprise richness, is a significant driver of the scaling of enterprise and population numbers. The wealth/poverty states of towns, measured as their enterprise dependency indices, modify the relationship between enterprise richness and population numbers. The development of plans for sustainable development without consideration of the quantitative orderliness of the demographic-economic-entrepreneurial nexus of this biosphere reserve and the factors that control it, would be dealing in 'a strategy of hope' rather than 'a strategy of reality'. Achieving the latter in a system with regularities, some of which that have existed over decades, is a huge challenge. Understanding of how productive knowledge can be leveraged, should be a key component in any strategy.

In: Advances in Environmental Research ISBN: 978-1-53615-009-4
Editor: Justin A. Daniels

Chapter 1

IODIDE MOBILITY, TRANSFORMATION AND BEHAVIOUR IN A SOIL-FUNGI-PLANT SYSTEM

Eva Duborská*[*], *Martin Urík and Marek Bujdoš
Institute of Laboratory Research on Geomaterials, Faculty of Natural Sciences, Comenius University in Bratislava, Bratislava, Slovakia

ABSTRACT

Iodine is an essential micronutrient for humans because it is necessary for the synthesis of thyroid hormones. The most abundant iodine species are iodide and iodate, from which iodide represents the most bioavailable form for hormone production. Despite the importance of iodine, only a few studies to date have dealt with its distribution and mobility in the environment. The primary source of iodine is soil. From soil, it is translocated into plants, where the upper levels of the food chain are enriched. The translocation of iodine from soil depends on various (bio)geochemical factors, that influence iodine's mobility by affecting the bioavailable fraction for primary producers. The main criteria for iodide availability are water solubility, sorption to soil components and

[*] Corresponding Author Email: duborska.eva@gmail.com.

microbial activity. The presence of soil microorganisms prevents or limits the plant's uptake of iodine by its microbial accumulation. Furthermore, microorganisms can metabolically transform bioaccumulated iodine into volatile derivatives, such as methyl-iodide, and, thus, alter the content of bioavailable iodide in soils. There are also several other biogeochemical factors in soil that contribute to the change of iodine speciation in soil. Some soil redox active constituents can efficiently transform iodate into iodide or vice versa. The same affect can be ascribed to the metabolic activity of plant roots or microorganisms. Therefore, we present in this chapter a review on the geochemical behaviour of iodide (and iodate) in soils. Most of the information provided is accompanied with our own experimental data on soil-plant-microbial interactions with iodine, which provide new insight into iodine mobility in soils, as well as its bioavailability to plants and filamentous fungi. Experimentally, bioaccumulation efficiency of both readily bioavailable water soluble iodine species and a stabilised (aged) iodine from soil, which had previously gone through various geochemical transformations including sorption onto soil particles, were evaluated. The accumulated amount of iodine, as well as its distribution in soils, was determined by various one-step or sequential extraction procedures, as well as a few other methods of analytical geochemistry. We also provide new information on factors affecting soil-to-plant transfer factor and iodine phytotoxicity. Biologically induced loss of iodine via volatilisation into the atmosphere, which is caused by the biochemical transformation of iodine into volatile species by plants and microscopic filamentous fungi, are also evaluated. This chapter therefore, highlights current knowledge on iodide mobility in a soil-plant system and provides a theoretical and experimental basis for better understanding of the geochemical behaviour of iodine in soils, including its availability in the food chain.

1. INTRODUCTION

There are several reasons for having a scientific interest in the research of iodide's environmental mobility and the factors affecting its transfer into biological systems. The first reason is its biological function in higher organisms. Iodine, a micronutrient for humans and animals, is required for the synthesis of hormones by the thyroid gland. There is an entire spectrum of diseases caused by iodine deficiency or excess, and some of them are listed in Table 1. For example, insufficient intake of iodine during pregnancy may cause abortions or cretinism in newborn and stillbirth (Khun and Čerňanský, 2011).

Table 1. List of major health consequences caused by iodine deficiency

	Health consequences
All ages	Goiter
	Hypothyroidism
	Increased susceptibility to radioiodine intake
Fetus	Spontaneous abortion
	Stillbirth
	Congenital anomalies (insufficient mental development
	Perinatal mortality
Newborns	Endemic cretinism
	Insufficient mental development
	Mortality
Children and adolescents	Decreased mental functions (dementia)
	Delayed mental development Hypothyroidism
	Iodine induced hyperthyroidism
Adults	Decreased mental functions (dementia) Hypothyroidism
	Iodine induced hyperthyroidism

Table 2. Iodine deficiency classification according to the International Council for Control of Iodine Deficiency Disorders (WHO, 2007)

Average urinal iodine content [$\mu g.dm^{-3}$]	**Classification**
< 20	Severe deficiency
20-49	Serious deficiency
50-99	Slight deficiency
100-199	Sufficient intake
200-299	Increased intake
>300	Excessive intake

The recommended daily intake for adults is approximately 150 μg, but this value may vary depending on the age, development stage, and physical load of the individual. The recommended daily intake for men, however, is lower than for women whose recommended daily intake is 200 μg or higher (WHO, 2007). The adult human body contains approximately 30 - 50 mg of iodine of which less than 30% is accumulated in the thyroid gland and in its hormones (Venturi et al., 2000). Another significant amount of iodine in the human body is found in the salivary glands (Venturi and Venturi, 2009) and in other extra-thyroidal tissues, yet its biological role in these regions is still unclear (Venturi et al., 2000). Approximately 115 μg of iodide is taken up by the thyroid gland daily, of which about only 75 μg is used for thyroid hormone synthesis and for thyroglobulin storage. The rest is returned to extracellular fluids. The thyroid gland stores approximately 10 mg as a supplement for the protection of the body from iodine deficiency. Absorbed iodine is excreted in urine, therefore, the content of urinal iodine is the most appropriate indicator of its intake (Greenspan, 2003). Thus, International Council for Control of Iodine Deficiency Disorders frequently uses the classification of regions according to the amount of iodine excreted in the urine by the local population (Table 2). Classifying regions by iodine content in soils is unsuitable due to its complex geochemistry in soils and different bioavailability rates.

To eliminate iodine deficiency disorders, iodination of table salt is the most commonly used remedy (WHO, 2014), but it is not sufficient in some cases. The primary source of iodine for humans is food. Most iodine contains seafood, meat, milk and eggs, or some plants such as strawberries or cranberries (Hejtmánková et al., 2005; Zimmermann et al., 2008). This is why some authors also recommend the fortification of agricultural crops using iodine containing organic fertilizers, e.g., iodine algae with a high iodine content, or iodine enriched irrigation water (Ren et al., 2008; Weng et al., 2014; Weng et al., 2013). Still, the consumption of nutritional supplements containing iodine may, in some cases, induce autoimmune thyroiditis (Zois et al., 2003).

The second reason, why the environmental mobility of iodide is a scientific concern is the potential exposure to radioactive iodine isotopes that are considered dangerous for human health. There are 37 known iodine isotopes of which only isotope ^{127}I is stable. From all of the known radioactive iodine isotopes, fourteen of them have a half-life longer than 10 minutes. Among the most widespread radioisotopes are ^{129}I with a half-life of 15.7 million years, and ^{131}I, with a half-life of about eight days (Hansen et al., 2011). Their ability to enter the food chain is depicted in Figure 1.

Radioactive isotopes of iodine are continuously emitted into the environment mainly by anthropogenic activity (nuclear weapons tests, radioactive waste and nuclear power plant disasters). It is estimated that around 60 kg of ^{129}I was released into the atmosphere during the atomic bomb tests, and approximately 2-6 kg in the Chernobyl disaster (Englund et al., 2010; Hou, 2009). Despite their dangerous effects, radioisotopes ^{126}I and ^{125}I are used in medicine for the diagnosis and treatment of thyroid diseases (Maxon et al., 1997; Shimura et al., 1997).

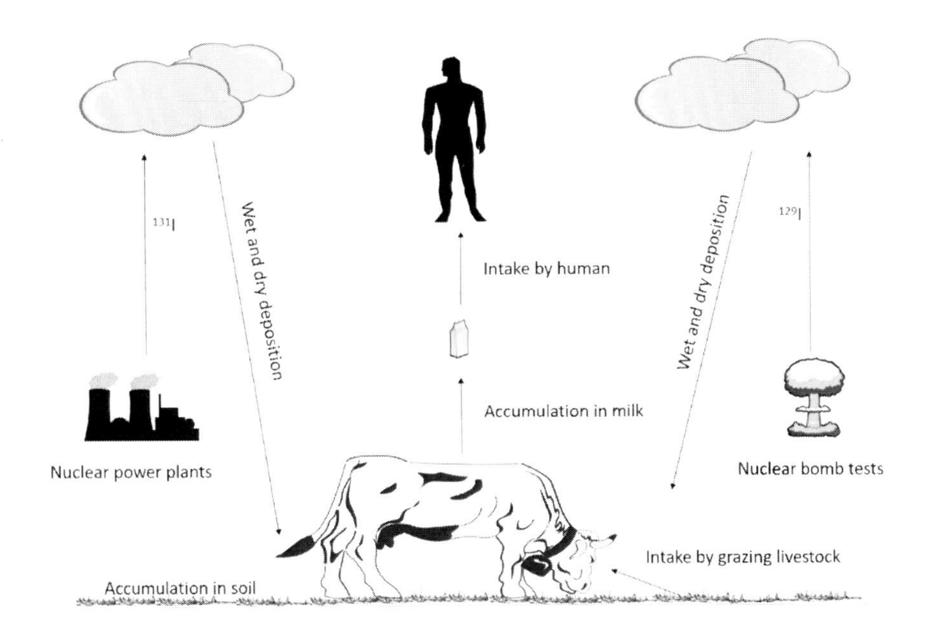

Figure 1. The pathway of radioactive iodine isotopes entering the food chain.

2. Factors Affecting Iodide Mobility in Soils

In a terrestrial environment, the dominant form of iodine is iodide (I^-), which may be converted under oxidative conditions into iodate (IO_3^-), the other dominant iodine species in the environment. In extreme acidic and oxidative environmental conditions, hypoiodous acid (HIO) and iodic acid (HIO_3) can also be stable in aqueous solutions, while the periodate ions (IO_4^-) are stable under oxidative and alkaline conditions. As highlighted in Figure 2, the redox conditions and pH of the environment have a significant impact on iodine speciation. This then affects iodine mobility in soils (Hu et al., 2009; Johnson, 2003; Muramatsu and Hans Wedepohl, 1998). Furthermore, due to the complexity of soil structures and involvement of biotic component, all iodine species may occur in a soil system simultaneously (Emerson et al., 2014; Korobova, 2010; Yamada et al., 1999).

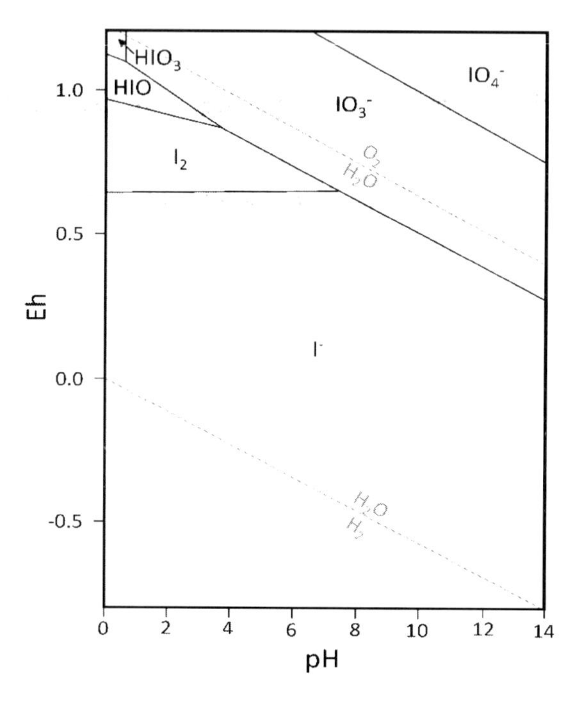

Figure 2. Eh-pH diagram for iodine inorganic species.

Because of the complex iodine geochemistry in soils and its high affinity to organic matter, clay minerals and iron, manganese and aluminum oxides (Hu et al., 2009), iodine can be found in several chemical forms, as well as in association with various soil components. The main iodine fractions in soils can be classified according to their mobility as (1) water soluble/leachable and (2) exchangeable iodine fraction, iodine fraction bound to (3) oxides or (4) organic matter, and (5) non-mobilisable iodine fraction. Leachable fraction, which consists mostly of weakly bound iodides, is the most mobile and bio-accessible form of iodine, whereas iodates and elemental iodine tend to bind onto oxides and organic matter, and are thus considered lesser mobile inorganic iodine species (Hansen et al., 2011).

Substances which have an anionic character, such as iodides are very poorly bound to soil particles and are easily leachable from the soil (Giles et al., 1974), weakly bounded anions are more mobile and bio-accessible in a given environment. However, as a result of their leachability, iodine deficiency may occur especially in areas which are periodically flooded or where rainfall is frequent (Podoba, 1962).

2.1. The Effect of Redox Conditions on Iodine Speciation in Soils

Redox soil properties are an important factor that affects iodine abundance, speciation and retention in soils (Yuita and Kihou, 2005). Generally, the total soil iodine concentration decreases in the soil profile with increasing depth, especially in soils where the surface layers have a higher redox potential. In wet or flooded soils, where reductive conditions prevail in the surface layers, soil iodine content increases with increasing depth, as well as the potential increase in redox. Therefore, it appears that in anoxic conditions, iodine bound to soil particles can be easily reduced and released into soil solutions (Yuita et al., 1991).

Oxidative and reductive transformations are also associated with the occurrence of various iodine species due to the significant changes of geochemical and environmental conditions. Yuita et al. (1991) examined the presence of iodine species under various redox conditions. In soils where oxidative conditions predominated, up to 86% of iodine was found as iodate and only 6% as iodide, while 4% was associated with organic matter and only 4% was identified as elemental iodine in a soil solution. However, after a change in soil condition to a reductive one, the iodate content significantly decreased to just 12%, which is probably the result of its partial reduction to iodide, whose abundance increased up to 87%. The amount of elemental iodine and organic iodine compounds also declined, however, the changes observed were not that significant.

Oxidation of iodide to iodate takes place in two steps. Firstly, iodide is oxidized to elemental iodine or hypoidous acid. These are very reactive species, which are subsequently oxidized to iodate (or bound to organic matter). The oxidative transformation of iodide to iodate can be catalysed in the presence of manganese oxides (δ-MnO_2) in acidic environments. The newly formed iodate is then bound to the manganese oxide surfaces. The oxidation rate of iodide increases with the concentration of manganese oxides and decreases with increasing pH of the medium (Fox et al., 2009).

In the presence of soil organic matter, e.g., humic substances, the oxidation of iodide to iodate is limited by the reaction of the intermediate product of elemental iodine with humic substances for forming organic iodine compounds (Fox et al., 2009; Gallard et al., 2009). The reduction of iodate also takes place under specific conditions, leading to the formation of elemental iodine and chemically unstable hypoiodous acid, which are then immobilised in the soil by binding to soil particles or organic matter (Otosaka et al., 2011).

2.2. The Impact of Soil pH on Iodine Mobility in Soils

Inorganic forms of iodine can be trapped in the soil on positively charged surfaces of iron, aluminum and manganese hydroxides, as well as on clay minerals, especially when the pH of the medium is below 6 (Dai et al., 2006). Mobility of iodine inorganic forms decreases with increasing soil pH. The optimum value for iodine binding is estimated at 3.7, but smaller amounts were also bound at pH 6.5 (Emerson et al., 2014).

Iodide and iodate are already completely dissociated at normal pH values, therefore they can be electrostatically drawn to various positively charged iron oxide surfaces. This binding is, however, more prominent under acidic conditions (Kaplan et al., 1999; Um et al., 2004). Yet, inorganic iodine sorption was also observed under slightly alkaline conditions (Shimamoto et al., 2010; Yoshida et al., 1992), and quite extraordinarily at pH 9.4 (Kaplan et al., 1999). The low pH in soils with a low organic matter content creates suitable conditions for iodate binding (Kaplan et al., 1999; Shimamoto et al., 2010; Um et al., 2004; Yoshida et al., 1992).

The pH value has a particularly significant influence on the course of iodide sorption in soils, since iodide as an anion has a low affinity to humic acids and clay minerals in neutral and moderate alkaline soils. The surface of both components is negatively charged in such a medium and therefore iodides remain mobile (Johnson, 2003). The pH value also affects iodine preferential sorption to organic matter fractions. It was observed, that at pH below 5.5, iodine binds preferentially to humic acids, however at pH > 6 they are immobilised more efficiently by fulvic acids (Hansen et al., 2011).

2.3. Iodide Sorption to Soil Components

As mentioned before, both redox conditions and soil pH significantly affect sportive immobilisation of iodide due to (1) its transformation to species with a significantly different affinity towards soil components, as well as (2) changes in the net charge of soil components' surfaces. Therefore, a more detailed analysis of sorption onto the most significant components of soil is presented in the following subsections.

Here, the term sorption represents the immobilisation of iodide and its species by natural components of soils or sediments (Yuita, 1992). In order to express the sorption rate and specific binding mechanisms, mathematical sorption models expressed as sorption isotherms and kinetics are frequently used in experimental studies. These models allowed us to evaluate the soil/sediment sorption capacity or iodide affinity to particular soil components directly.

The most frequently used sorption models for sorption evaluation are the Freundlich (1) and Langmuir (2) adsorption models:

$$S = K_f C_i^n \tag{1}$$

where S [mg.kg^{-1}] is the amount of adsorbed iodide and its other species per soil unit mass; C_i [mg.L^{-1}] is the remaining concentration of iodine in the solution after an equilibrium state was reached during sorption; K_f [L.kg^{-1}] is Freundlich sorption coefficient which expresses the sorption capacity of the sorbent when the equilibrium concentration of sorbate in the solution is unitary; and n (dimensionless variable) is Freundlich's sorption exponent which expresses the heterogeneity of the sorbent sorption positions and the course of the sorption isotherm.

$$S = \frac{K_L S_{max} C_i}{1+K_L C_i} \tag{2}$$

where S_{max} [mg.kg^{-1}] represents the maximum amount of sorbate that can adsorb to the sorbent's unitary weight; and K_L [L.mg^{-1}] expresses the sorbate affinity for the sorbent (Hiller et al., 2012; Hong et al., 2012).

The suitability of these models for the expression of iodide sorption is confirmed by several works where the Freundlich (Emerson et al., 2014; Fox et al., 2009; Gallard et al., 2009; Yuita, 1992) and Langmuir (Dai et al., 2004; Gallard et al., 2009) adsorption models were used. In the event that the initial iodide concentrations are very low or a narrow concentration range is applied in the studied sorption system, linear sorption models are also suitable (Emerson et al., 2014; Sheppard et al., 1996; Sheppard, 2003).

The factors that affect the mechanism and nature of sorption are very complex and include (1) the properties of the sorbate, e.g., its chemical properties and forms of occurrence – speciation, (2) the soil/soil component chemical and physical properties, e.g., ion exchange capacity and surface net-charge, and (3) the environmental conditions under which the sorption process takes place, e.g., redox conditions, soil reaction, biological processes and others (Brusseau and Chorover, 2006).

The amount and quality of soil components to which iodine has a high affinity also have a high influence on iodine sorption in the soil system. Iodide is presumably best sorbed in soils with a high organic matter content and high abundance of oxohydroxides of various metals (especially iron, aluminum and manganese). Some authors concluded that calcium compounds can also increase iodine binding (Sheppard and Thibault, 1992; Toyohara et al., 2002). In general, iodine sorption is a very slow process, as sorption equilibrium occurs only after approximately eight days (Emerson et al., 2014).

2.3.1. Sorption to Clay Minerals

Soils rich in clay minerals usually exhibit a higher iodine content (Shinonaga et al., 2001). Both iodide and iodate sorption is achieved through binding on positively charged components, whose quantity in soils is limited. Clay minerals typically have a small anionic exchange capacity, due to the presence of positively charged structural loops. Iodine anions can occupy these positions by ion exchange mechanisms with hydroxide anions, especially at a higher pH. Thus, small amounts of iodine can bind to positively charged edges of 1: 2 type clay minerals in which one octahedral layer is surrounded by two tetrahedral layers, such as montmorillonite and illite, as well as clay minerals of the 1: 1 type containing one tetrahedral and one octahedral layer such as kaolinite (Hu et al., 2009; Christiansen and Carlsen, 1989; Johnson, 2003).

The presence of some dissolved calcium salts reduces the sorption capacity of montmorillonite and bentonite for iodine (Raja and Babcock, 1961). Clay minerals in soil also play an important role in the prevention of iodine volatilisation the soil into the atmosphere. Its release is likely dependent on the redox properties of the clay. However, clay minerals are not considered to be important sorbents of iodine in soil (Hu et al., 2009), but indirectly affect its mobility. Iodate can be reduced by the action of divalent iron, which may be present in clay minerals, thereby creating weakly bound iodine forms (Hu et al., 2005).

2.3.2. Sorption to Iron, Aluminum and Manganese Oxyhydroxides and Other Mineral Phases

Sorption of iodine to minerals also depends on its speciation. Magnetite (Fe_3O_4) was reported to be capable of binding iodide but not iodate. Conversely, iodate is sorbed onto biotite [$K(Mg,Fe)_3(AlSi_3O_{10})(F,OH)_2$], yet iodide remains in the solution. During the adsorption process, some minerals are able to change the redox state of the environment and thereby influence its speciation and sorption. For example, in the presence of pyrite (FeS_2), iodate (which is not likely to bind) is reduced to I_2, which is able to form a sorption complex (Fuhrmann et al., 1998). A similar example is the oxidation of iodide and its subsequent sorption to MnO_2 as mentioned above.

Because of iodine's ability to bind to metal oxides, there is a close relationship between iodine concentration and the iron, aluminum and manganese content in soil. This type of sorption takes place predominantly under oxidative conditions. Under reducing conditions, the dissolution of metal sesquioxides causes the release of the adsorbed iodine into the soil solution (Um et al., 2004; Yuita et al., 1991). In acidic soils with low organic matter content, iodate is usually sorbed more effectively than iodide, which may be due to the fact that iodate is able to create chemical bonds with iron oxides by replacing hydroxyl groups on their surfaces (Shimamoto et al., 2010; Um et al., 2004).

Minerals such as hematite (Fe_2O_3) and several aluminum oxides can directly adsorb iodide from aqueous solutions at pH up to 8 (Whitehead, 1974). Iodate can be strongly bound to hematite to pH 8, however, iodide is sorbed poorly (Couture and Seitz, 1983).

The anion exchange capacity also plays an important role in the sorption of iodide onto mineral components. The anion exchange capacity determines (a) the presence of exposed cations in the edges of the crystalline structure, or (b) the ionization of groups in colloids such as $Al(OH)_3$ and $Fe(OH)_3$ or the isomorphic exchange of Ti^{4+} for Fe^{3+} in iron oxides. Generally, anion exchange capacity increases with a decrease in the soil's pH value. Soils with a higher sesquioxide content may have a higher anionic exchange capacity, which enables the binding of iodide anions (Čurlík and Jurkovič, 2012).

2.3.3. Sorption to Soil Organic Matter

Iodine is able to accumulate in soil organic matter whereas humic substances are its largest reservoir (Dai et al., 2009). Dead plants incorporated into soil with iodine of atmospheric origin are at the same time a significant source in soil (Yoshida, 1999). In soils and sediments from 30 to 55% of stable iodine, isotope (^{127}I) is associated with soil organic matter, while 42 to 60% of the long-lived radioactive isotope ^{129}I associated with soil organic matter indicates its higher affinity for organic matter (Hansen et al., 2011).

The iodine content of soil decreases with a decreasing amount of organic matter in a vertical profile with increasing depth (Aldahan et al., 2007; López-Gutiérrez et al., 2004). Iodine sorption to organic matter occurs through the mechanism of slow diffusion of iodine into micropores and cavities in the structure of organic matter (Sheppard and Thibault, 1992). Humines can act as an electron donor and acceptor in soil. Therefore, they can oxidize iodide and reduce iodate to I_2 or HOI, which easily react with soil organic matter. The sorption mechanism of iodate is therefore its reduction by humic substances to the reactive species I_2 and HOI, as well as their subsequent chemisorption to organic matter by the mechanism of electrophilic substitution which primarily takes place on the aromatic ring of humic components (Reiller et al., 2006). The presence of amino groups in humic substances can also increase iodine sorption capacity in the soil (Čurlík and Jurkovič, 2012).

Iodide sorption takes place by similar mechanisms. First, iodide is oxidized to HOI and I_2, which subsequently react with organic matter. According to some authors, binding of iodine to organic matter takes place by electrostatic attractions or other physical mechanisms as well (Sheppard and Thibault, 1992). Both humic and fulvic acids can be active sorbents depending on the pH conditions of the soil (Hansen et al., 2011). Yamada et al. (1999) observed iodine adsorption onto soil with a pH range of 4.8-5.3 when 63% of iodine was bound to humic acids and 20% to fulvic acids.

3. Biotic Factors Affecting Transformation of Iodide

3.1. Microbial Reduction

In their effort to explain how microorganisms affect iodide and iodate distribution in ocean waters, Tsunogai and Sase (1969) demonstrated that iodate can be reduced aerobically to iodide by nitrate-reducing marine bacteria, as well as by nitrate reductase extracted from *Escherichia coli*. Since nitrate reductase is more commonly present in phytoplankton, various authors assumed that phytoplankton plays a significant role in the global geochemistry of iodine. This hypothesis was successfully incorporated into the general view of the iodine geochemical cycle and was supported by an observed linkage between iodide formation and nitrate uptake. Therefore, even though both abiotic and biotic processes had been accounted for iodate reduction to iodide, there was more emphasis on the biotic transformation.

Truesdale (1978) suggested, however, that the iodate-reducing substances present in terrestrial run-off are more likely responsible for maintaining the iodide/iodate ratio in temperate oceans. They discovered that there were only negligible temporal changes in iodate concentrations during seasonal nutrient cycling in shell-waters. The correlation between primary production and iodate concentrations was also challenged by Tian et al. (1996) who suggested that the iodate reduction to iodide in surface waters was essentially regulated by recycled (regenerated) production and therefore, primary production by algae affects iodine speciation negligibly. This was further supported by Bluhm et al. (2010), who discovered that iodate reduction resulted from changes in cell membrane permeability in the senescent phase of algal growth. As Taurog et al. (1966) demonstrated, iodate could be rapidly reduced to iodide non-enzymatically in the presence of glutathione, therefore, we may conclude that the release of glutathione or other reduced internal sulfur containing organic molecules during cell apoptosis is beneficial for iodide formation.

Truesdale and Bailey (2002) concluded that iodate and nitrate enzymatic reduction is most likely independent from one another. The role of nitrate reductase in iodate reduction was thus challenged by Mok et al. (2018). They demonstrated that iodate reduction by *Shewanella oneidensis* was not competitively inhibited by the presence of nitrate; and they suggested that iodate is reduced via an unknown mechanism. Kengen et al. (1999), however, isolated iron, molybdenum and selenium containing a periplasmic oxygen-sensitive chlorate reducing enzyme, which exhibits both nitrate and iodate reductase activity.

As already indicated, microbial reduction of other halogen oxyanions, such as chlorate and bromate, has been demonstrated under low oxygen concentrations (Davidson et al., 2011; Lai et al., 2018). Similarly, it was expected that microbial reduction of iodate to iodide could therefore be favored in anaerobic conditions. Councell et al. (1997) demonstrated that in the presence of iodate as the exclusive electron-acceptor, the sulfate-reducing bacterium *Desulfovibrio desulfuricans* was capable of reducing almost 96% of 100 μmol. L^{-1} iodate in a nutrient-free medium. Similarly, *Shewanella putrefaciens*, a facultative anaerobe, showed reasonably comparable iodate reduction efficiency in synthetic seawater, however, Farrenkopf et al. (1997) suggest that there is an intermediate step in reduction process, where the iodine forms bonds with organic carbon. These structures are then abiotically or microbially hydrolyzed and iodide is released (Truesdale and Luther, 1995). Organic carbon bonds with iodine are relatively weak and, as demonstrated by Wong and Cheng (2001), the photochemical decomposition of dissolved organic iodine by irradiation strongly correlated with iodide production. In addition, abiotic reduction of iodate by soluble ferrous iron, sulfide and iron monosulfide has been demonstrated (Councell et al., 1997; Jia-Zhong and Whitfield, 1986). Therefore, microbial contribution in iodate reduction to iodide under various environmental conditions of aquatic or terrestrial environments is still unclear. However, various experiments indicate that the iodate reduction as well as its efficiency can be easily enhanced by microbially produced metabolites or biotic, direct, involvement of microorganisms via enzymatic transformation of iodate.

3.2. Iodide Biomethylation

Bell et al. (2002) calculated that the ocean accounts for 70% of the annual flux of methyl iodide into the atmosphere, while rice paddies contribute to a global budget of methyl-iodide by 24%. The other sizeable sources of methyl-iodide are natural wetlands and peatlands (Dimmer et al., 2001), while global emissions of methyl-iodide by biomass burning are less significant (Blake et al., 1996). It was suggested that besides the photochemical degradation of refractory organic material or some other photochemical reactions (Moore, 1994), the biological activity of microorganisms and algae also directly contribute to iodine transfer into the atmosphere in the form of volatile methyl-iodide (Stemmler et al., 2014). This biotically driven transformation is executed via specific reaction pathways usually involving enzymatic conversion of iodide. Thayer (2002) suggests that there are three possible pathways for iodide biomethylation. The first possible transformation pathway is iodide reaction with β-dimethylsulphoniopropionate. Dimethylsulphonio-propionate is an algal osmoprotectant which is readily degradable by chemoheterotrophic bacteria to dimethylsulfide (Yoch, 2002). However, Hu and Moore (1996) demonstrated that iodide methylation using dimethylsulphoniopropionate is not a thermodynamically feasible reaction since its free energy charge is positive and, thus therefore, its estimated contribution to the global oceanic production of methyl-iodide is negligible (less than 1%). The possible second mechanism of methyl-iodide formation is an enzymatic transformation by vanadium-dependent haloperoxidases. Moore et al. (1996) suggested that haloperoxidase did not contribute to the formation of monohalogenated halocarbons, such as methyl-iodide. Iodoperoxidase, identified in various marine diatoms and macrophytes (Colin et al., 2003), more likely facilitates the formation of polyhalogenated organic compounds, mainly CH_2I_2 (Carpenter et al., 2000; Moore et al., 1996). Fuse et al. (2003) isolated two strains of marine bacteria, closely related to *Roseovarius tolerans*, which were capable of producing considerable amounts of free iodine and iodoform from iodide, while the production of methyl-iodide was negligible.

Finally, the third proposed mechanism by Thayer (2002) suggests direct transformation of iodide into methylated species via enzymatic mechanism proposed by Challenger (1951). Via this methylation pathway, transformation of inorganic iodide into methylated compounds involves the transferase mediated nucleophilic attack of iodide at the electrophilic CH_3^+S site of the S-adenosyl-L-methionine methyl donor (Itoh et al., 1997). This is supported by Amachi et al. (2001) who used cell extracts of *Rhizobium* sp. *in vivo* to demonstrate that iodide methylation is S-adenosylmethionine dependent. In this study, it was also demonstrated that methyl-iodide producing ability is widespread among marine and terrestrial bacteria (e.g., *Variovorax* sp., *Photobacterium leiognathi*), while anaerobic microorganisms (e.g., methanogens) were incapable of producing methyl-iodide. Furthermore, Manley and Cuesta (1997) have shown that marine phytoplankton's methyl iodide production under laboratory conditions is not very significant. Butler et al. (1981) hypothesized that even when produced by phytoplankton, methyl-iodide is only a short-lived intermediate which is converted into methyl-chloride. Nevertheless, there is an estimation that cyanobacterium *Prochlorococcus* alone contributes to the annual global flux of methyl-iodide by $4.3x10^9$ mol (Smythe-Wright et al., 2006). However, the other dominant primary producer *Synechococcus* did not produce any methylated volatile derivatives in laboratory based experiments (Hughes et al., 2011).

Since the reported sweater concentrations of methyl-iodide within and outside the beds of common species of northern temperate marine macroalgae, including *Laminaria digitata*, ranged from 2.7 to 5.1 $ng.L^{-1}$, Nightingale et al. (1995) suggested that macroalgae are not an important source of this monohalogenated iodine compound in the atmosphere. Similarly, Manley and Dastoor (1987) estimated that the methyl-iodide production by kelp accounted for less than 0.1% of its annual oceanic global production. They found that temperate kelp *Macrocystis pyrifera* released 98.6 pmol methyl-iodide.g^{-1} of wet algal weight daily. Also, Laturnus et al. (1998) reported that the mean value of daily methyl-iodide production by various Antarctic macroalgae was only 1.68 pmol.g^{-1} of wet algal weight. In conclusion, this highlighted only a minor role of macroalgae in global methyl-iodide production.

As mentioned previously, rice paddies also contribute significantly to the global budget of methyl-iodide (Redeker et al., 2000). In their early work, Muramatsu and Yoshida (1995) suggested that methyl-iodide is produced in flooded soil with rice plants under ananerobic conditions with a possible contribution from both microbial and root activity. More importantly, they speculate that methyl-iodide is emitted into the atmosphere through plant shoots, similarly to the mechanism of methane release from rice-fields (Aulakh et al., 2001). However, Amachi et al. (2003) discovered that anaerobic conditions did not facilitate production of methyl-iodide, even when various electron donors were introduced into soil paddies during incubation in order to stimulate activity of anaerobic soil microorganisms. Furthermore, they suggested that biovolatilisation of iodine is not confined to any particular bacterial group and, thus, transformation of iodide into volatile methyl-iodide should be a global phenomenon. However, iodide biomethylation in terrestrial ecosystems is not an exclusive characteristic of bacteria. Ban-nai et al. (2006) identified eleven filamentous fungal strains, including strains of basidiomycete, ascomycetes and fungi imperfecti, which were capable of emitting methyl-iodide. No other gaseous organic iodine compounds were identified. Ban-nai et al. (2006) also suggested that fungi have a comparable volatilisation potential to that of soil bacteria. This is supported by Harper (1985), who assessed methyl-iodide production by common wood-rotting fungus *Phellinus pomaceus* with a highly efficient (over 60% of 1 mM iodide in four weeks) biologically induced conversion of iodide into methyl-iodide. Redeker et al. (2004b) indicated that the daily iodide emission rate via volatile methyl-iodide significantly varies among species and morphotypes of ectomycorrhizal fungi within a range from 0.15 to 30 $\mu g.g^{-1}$ of fungal dry biomass at 0.2 $mmol.L^{-1}$ iodide concentration.

Still, Redeker et al. (2004a) and Redeker and Cicerone (2004) concluded that rice plants alone are capable of producing methyl-iodide, especially during the early season when the plants rapidly increase in biomass and height, with a minor role of emissions from water or soil columns. They also suggest that methyliodide production is an iodide detoxification mechanism because, compared to other halo-ions, the high emission rate of methyl-iodide by rice implies the presence of methyl transferase with a high affinity to iodide and high catalytic efficiency. Furthermore, it was reported that iodide is the preferred substrate for methylation compared to other halogens (Harper, 1985).

The enzyme responsible for iodide methylation in cabbage leaves (*Brassica oleracea*) was identified as halide/bisulfide methyltransferase with pH optimum for iodide methylation in the range from 5.5 to 7.0. This enzyme catalyzes the S-adenosyl-L-methionine-dependent methylation of the halides iodide, bromide, and chloride to monohalogenated methanes (Attieh et al., 1995). Further assays indicated that the herbaceous plant species from 44 families and 33 orders were capable of producing readily detectable amounts of methyliodide (Saini et al., 1995), therefore, it appears that iodide methylation is widespread among higher plants.

3.3. Iodide Biooxidation

Although Yasuo and Shizuo (1963) discovered that the coexistence of iodate and iodide led to some molecular iodine formation under laboratory conditions, they suggest that it is very unlikely that a detectable amount of molecular iodine is produced from iodide in natural sea water without being irradiated by solar light. However, Wong (1982) suggests that in seawater molecular iodine is unstable and rapidly hydrolyzed into hypoiodous acid. Concurrently to abiotic oxidation, the biologically catalyzed oxidation of iodide was proposed.

Gozlan and Margalith (1973), identified the iodide-oxidizing Gram negative bacterium *Pseudomonas iodooxidans,* based on their report on microbially induced liberation of elemental iodine from seawater that led to death of the Mugil fish. They suggested that this strain was capable of oxidizing iodide by an extracellular enzymatic system in the presence of microbially generated hydrogen peroxide. In conjunction with iodide-oxidizing strains' high resistance to elemental iodine, microbially induced oxidation to iodine inhibits the growth of other microorganisms and provides an ecological advantage for iodide-oxidizers (Zhao et al., 2013), especially in iodine-rich environments. Thus, iodide-oxidizing bacteria are more likely to prevail in brine water, while occurring in natural seawaters and terrestrial soils only after significant iodination (Amachi et al., 2005b).

Iodide capability to scavenge aqueous hydrogen peroxide, as well as gaseous ozone, hydroxyl radicals and superoxide, led to the hypothesis that (bioaccumulated) iodide serves as an apoplastic antioxidant which is mobilized during oxidative stress (Küpper et al., 2008). However, Shah and Aust (1993a) suggested that lignin peroxidase of *P. chrysosporium* catalyzes oxidation of iodide to iodide radical, which serves as a mediator for the oxidation of organic acids.

A variety of enzymes' involvement in iodide oxidation has been reported, e.g., vanadium haloperoxidase from three different species of Laminariaceae (Almeida et al., 2001), extracellular lignin peroxidases of the white-rot fungus *Phanerochaete chrysosporium* (Farhangrazi et al., 1992; Shah and Aust, 1993b), lactoperoxidases (Furtmüller et al., 2002; Maguire and Dunford, 1972) and fungal laccases isolated from *Myceliophthom thermophile* (Xu, 1996).

Oxidation of iodide to hypoiodous acid was found to be an essential step for its bioaccumulation by the marine microalga *Isochrysis* sp. (Van Bergeijk et al., 2013). Küpper et al. (1998) suggested a mechanism for iodine bioaccumulation in *Laminaria*, where extracellular (apoplastic) haloperoxidase catalyzes iodine bioaccumulation, since synthesized lipophilic products of iodide catalyzed oxidation (hypoidous acid and molecular iodine) diffuse freely through plasma membranes. Similar iodide transformation prior to its uptake was also proposed by Amachi et al.

(2007) for marine Flavobacteriaceae bacterium strain C-21 which phylogenetically relates to *Arenibacter troitsensis*. They also indicated that hydrogen peroxide plays a key role in iodine uptake.

Iodide oxidation by laccase in soils is clearly involved in soil organic matter iodination (Seki et al., 2013). Enzymatic oxidative conversion of iodide prior to its binding was also highlighted by Lusa et al. (2015) who supported this hypothesis by finding a reasonably positive correlation between peroxidase activities in the surface of moss, subsurface of peat and the gyttja layers of a boreal bog with iodide batch K_d values.

Ball et al. (2010) suggested that the emission of molecular iodine from seaweed is also associated with iodide oxidation involving halo-peroxidases. This is most likely to counter an internal oxidative stress response resulting from exposure to air or oligoguluronates elicitation. Substantial amounts of molecular iodine are also produced from the algal kelp *L. digitata* via direct reaction between iodide and ozone (Palmer et al., 2005), and from *L. saccharina*, *Desmarestia antarctica* and *Neuroglossum ligulatum* after exposure to UV radiation (Laturnus et al., 2004).

3.4. Case Study: Fungal Bioaccumulation and Biovolatilisation of Iodide and Iodate

3.4.1. Introduction

Iodine is considered one of the less abundant elements in the environment. Soils represent an important source of iodide in terrestrial systems although the total amount of iodine is only 3 $mg.kg^{-1}$ on average (Johnson, 2003). Soil iodine is mostly bound to soil components, which depend on several environmental factors. Mobile iodine species, especially iodide, can be taken up by plants or other soil biota and enter the food chain. To date, biochemical mechanisms of iodide accumulation have been extensively studied only for mammals (Taurog et al., 1947) and brown algae (Küpper et al., 2013; Küpper et al., 1998).

Due to biogeochemical transformations of iodine in soil, which are part of its natural biogeochemical cycle (Amachi et al., 2001; Fuge and

Johnson, 1986), the iodine loss via (bio)volatilisation in the form of elemental iodine or organic gaseous iodinated substances such as methyl iodide (CH_3I), which is a result of the metabolic activity of soil microbiota, can limit its transfer to plants (Muramatsu et al., 2004). These mechanisms also significantly enhance its environmental mobility. Iodine volatilisation from soils was reported by Whitehead (1981), while the degree of such volatilisation efficiency decreases after soil sterilization (Amachi et al., 2003). This indicates the significance of soil microorganism's contribution to iodine volatilisation from terrestrial systems.

The marine environment represents the main contribution of the production of volatile iodine species. Additionally, terrestrial ecosystems are also considered important biogenic sources of volatile iodine species in the global biogeochemical cycle of iodine. Several fungal species have also been reported to possess halide methylating abilities (Ban-nai et al., 2006; Harper and Kennedy, 1986; Saini et al., 1995). Several terrestrial (e.g., *Rhizobium* sp., *Varivorax* sp.) and marine bacteria (*Alteromonas macleodii*, *Photobacterium leiognathi*, *Vibrio splendidus*) are capable of iodine methylating activities as well (Amachi et al., 2001; Amachi et al., 2005a). Methyl iodide production was also observed by ectomycorrhyzal fungi, *Cenococcum geophilum*, *Hebeloma crustuliniforme*, *Inocybe maculata* and *Laccaria lacata*, as well as by Basidiomycota *Lentinula edodes* (Redeker et al., 2004b). The same species also showed iodine accumulating and volatilising abilities (Ban-nai et al., 2006).

In the atmosphere, methyl iodide is photolyzed and produces elemental iodine atoms. These atoms are coupled with atmospheric ozone, H_xO_y and NO_x, resulting in inorganic iodine compounds such as IO, HOI, $IONO_2$ and I_2, which are subsequently transferred back to terrestrial and marine ecosystems (Chameides and Davis, 1980). As a result of these reactions tropospheric O_3 is depleted (Davis et al., 1996; Solomon et al., 1994). To date, there are only a few studies that provide information on iodide bioaccumulation and biovolatilisation ability of microscopic filamentous fungi. This inspired us to study and evaluate iodide and iodate and compare the mechanisms and the fungal bioaccumulation and volatilisation of these two most common iodine species in the environment in order to provide more information on iodine biogeochemistry and the effect of the interaction between iodine species and filamentous fungi on their mobility in the environment.

3.4.2. Materials and Methods

Five fungal strains: *Cladosporium cladosporioides* (G-3), *Aspergillus clavatus* (G-119), *Penicillium citrinum* (G-138), *Fusarium oxysporum* (G-93) and *Alternaria alternata* (G-97) were isolated from various agricultural soils from Slovakia by the Department of Mycology and Physiology of the Institute of Botany at the Slovak Academy of Sciences, and are currently deposited at the same institute. The fungal colonies were cultivated and maintained on Sabouraud agar slants (4% w/v) at 4°C. All isolates were classified to the genus or species level based on colony morphology, shape and appearance, colour and microscopic characteristics (mycelium septation, shape diameter and conidia texture) according to Nelson et al. (1983), Samson and Frisvad (2004), St-Germain and Summerbell (1996) and Pitt and Hocking (2009).

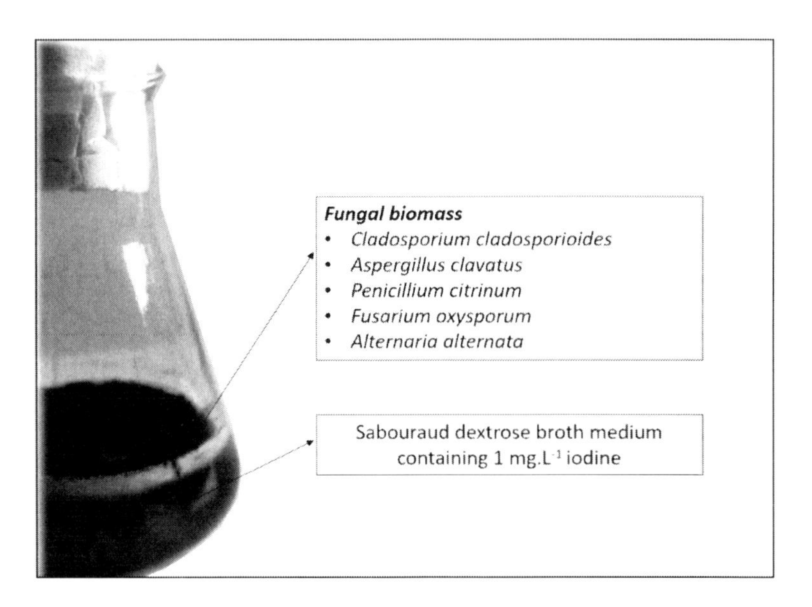

Figure 3. Experimental design for the evaluation of fungal contribution on iodide and iodate accumulation and volatilisation.

Iodide (I^-) and iodate (IO_3^-) stock solutions containing 10 mg.L^{-1} iodine were prepared by dissolving KI and KIO_3 (p.a. Centralchem, Slovakia) in deionized water under aseptic conditions.

Experiments for the evaluation of bioaccumulation were performed in 100 mL Erlenmeyer flasks containing a 45 mL Sabouraud dextrose broth medium (3% w/v) (HiMedia, Mumbai, India). The growth medium was sterilized in an autoclave at 120°C for 15 min before inoculation of the fungus. Seven day-old fungal culture suspensions were prepared and diluted in 5 mL of 10 mg.L^{-1} iodide and iodate solutions and were transferred to the growth medium under aseptic conditions. The final iodine content in the growth media was 1 mg.L^{-1}. The experimental design is illustrated in Figure 3. The cultures were incubated in the dark for 25 days. The control treatments without fungus were conducted in a similar manner to evaluate possible iodine volatilisation during the cultivation period. For each experimental condition triplicate parallel experiments were performed. After 25 days, the biomass was separated from the growth medium, rinsed with deionized water, dried at 25°C and weighed. The spent growth medium was filtered through 0.45 μm cellulose (MCE) membrane filter, pH was measured.

Total iodine content in the spent growth medium, water solutions and biomass extracts was determined by ICP-MS (Perkin Elmer Elan 6000, USA) using Te as an internal standard. Before iodine determination, 2 mL of 25% tetramethylammoniumhydroxide (TMAH) (Alfa Aesar, Germany) was added to the culture media and water solutions and diluted to 50 mL with deionized water in order to obtain 1% w/v background TMAH concentration for analysis. For total iodine determination in fungal biomass, extraction with TMAH was required. The dry biomass was added to 15 mL centrifuge tubes with 2 mL 25% w/v TMAH and placed in a dry bath incubator (MK-20, Hangzhou Allsheng Instruments Co., Ltd, China) for 4 hours at a temperature of 70°C. The samples were shaken manually approximately every 30 min. After 4-hours of incubation the samples were diluted to 50 mL with deionized water and centrifuged (CM-6MT, Sky Line, ELMI, USA) at 2300 rpm for 30 min. The supernatant was passed through a 0.45 μm pore sized MCE membrane filter. The efficiency of the extraction procedure was verified by analysing BCR certified reference material (Community Bureau of Reference certified reference materials) No. 129 (Hay powder) and GBW07405 (Stream sediment) acquired from the National Research Centre for Certified Reference Materials in China). The determined values of iodine corresponded appropriately with its certified values within the uncertainties.

3.4.3. Results and Discussion

Changes in the culture medium's pH and biomass weight are the most relevant fungal characteristics that indicate differences in physiological response to the effects of iodine species during cultivation. These are the most often used growth parameters for toxicity evaluation (El-Sayed, 2015). Changes in pH are usually omitted in papers dealing with the effects on microorganism of potentially toxic elements. The differences in pH reflect the organism's struggle with the efficient uptake of organic resources because the ATP-driven proton pump located in the membrane is responsible for maintaining the electrochemical proton gradient necessary for nutrient uptake (Manavathu et al., 1999). Our experimental data indicate that the effects of iodide and iodate on selected fungal strains did not significantly differ in most cases, as demonstrated in Figure 4. The average final pH of most fungal strains had an approximating value of 8. There was only a determined statistically significant difference for the strain *C. cladosporioides* (for $p < 0.05$) between the iodide and iodate treatments.

Some of our previous research on microscopic filamentous fungi (Urík et al., 2016) indicates that the lower final pH reflects the prolonged *lag* phase necessary for the fungi to accommodate for stress conditions. Thus, we can assume that iodate has higher toxicity for *C. cladosporioides* than iodide since the iodate treatment final pH was lower. On the other hand, Figure 5 indicates that the biomass production of strain *C. cladosporioides* was enhanced significantly (for $p < 0.05$) in the presence of iodate. Our results also suggest that no statistically significant difference in all other fungal strains' growth inhibition was induced either by iodide or iodate. Therefore, we concluded from the selected fungal strains that the only iodine sensitive strain was *C. cladosporioides* where iodate prolonged the *lag* phase and iodide decreased the dry biomass weight after cultivation.

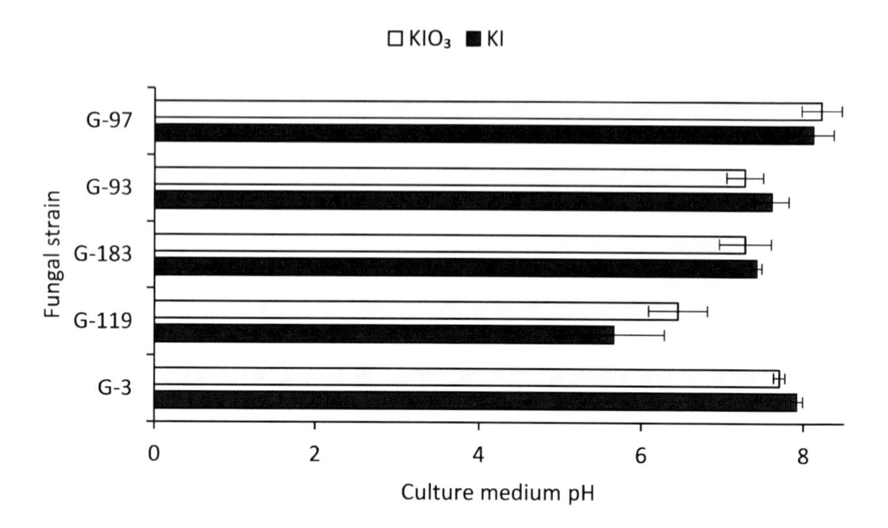

Figure 4. Changes in pH values of culture media after 25-day cultivation of selected fungal strains. The initial pH value was 5.6 (G-3 - *Cladosporium cladosporioides*; G-119 - *Aspergillus clavatus*; G-183 - *Penicillium citrinum*; G-93 - *Fusarium oxysporum*; G-97 - *Alternaria alternata*).

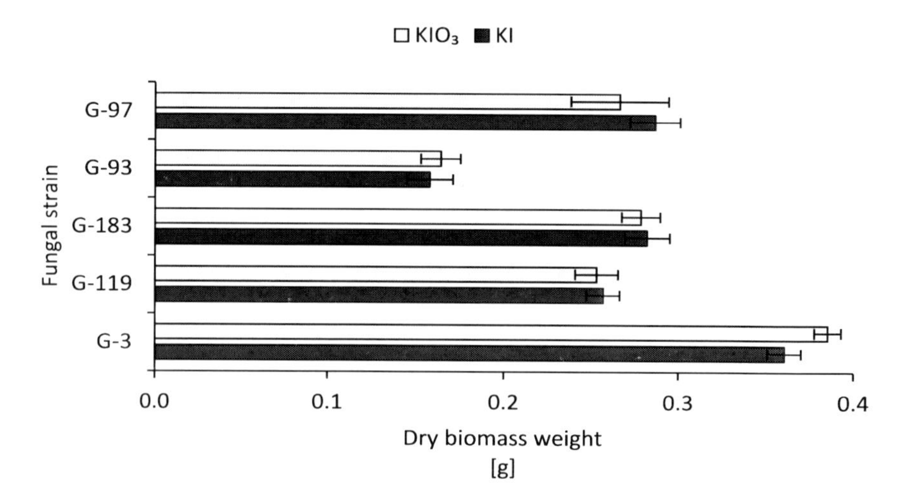

Figure 5. Dry biomass weight after the 25-day cultivation of the selected fungal strains (G-3 - *Cladosporium cladosporioides*; G-119 - *Aspergillus clavatus*; G-183 - *Penicillium citrinum*; G-93 - *Fusarium oxysporum*; G-97 - *Alternaria alternata*).

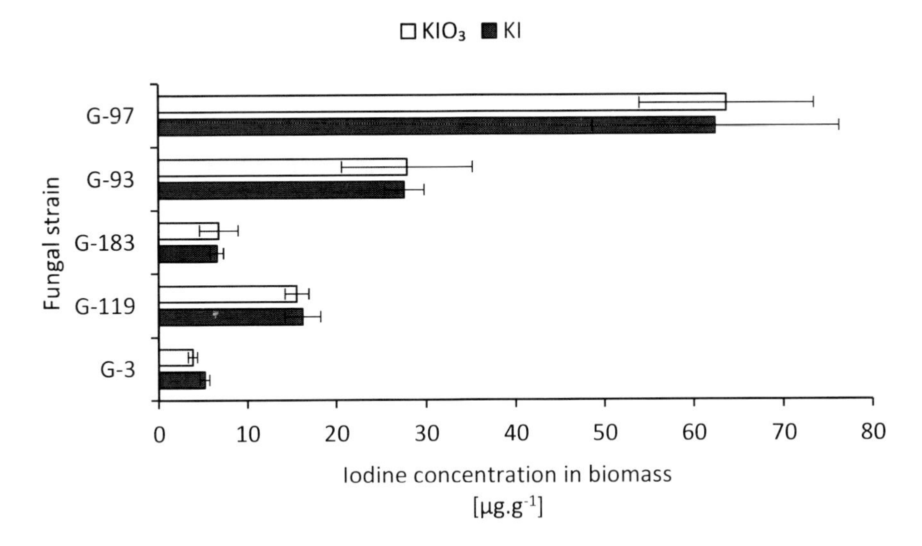

Figure 6. Iodine concentration in dry weight of biomass after 25-day cultivation. (G-3 - *Cladosporium cladosporioides*; G-119 - *Aspergillus clavatus*; G-183 - *Penicillium citrinum*; G-93 - *Fusarium oxysporum*; G-97 - *Alternaria alternata*).

The apparent bioaccumulation efficiency of iodate and iodide is shown in Figure 6. The only statistical difference in bioaccumulation efficiency of the two iodine species was observed for *C. cladosporioides*. This fungal strain accumulated 3.9 and 5.2 $\mu g.g^{-1}$ of iodine in the iodide and iodate treatments, respectively. The highest average iodine concentration - up to 63.7 $\mu g.g^{-1}$ in biomass was accumulated by *A. alternata* with no statistically significant difference in iodide and iodate bioaccumulation. Regarding the comparison of iodide and iodate accumulation efficiency, all other fungal species behaved similarly to *A. alternata*.

The specific uptake of iodine by the anaerobic microbial communities was estimated by Amachi et al. (2010) in the range of 0.71-2.0 $\mu g.g^{-1}$ dry weight of biomass, which is significantly lower than our *A. alternata*'s bioaccumulation efficiency. However, the aerobic strain of *Bacillus subtilis* 168 (MacLean et al., 2004) was reported to accumulate up to 152 $\mu g.g^{-1}$ iodide in biomass. The fact, that aerobic microorganisms have a significantly higher iodine accumulation capacity was also highlighted by Amachi et al. (2005a) with the aerobic marine bacterium *Arenibacter troitsensis*, which accumulated 55 $\mu g.g^{-1}$ iodide from the initial iodide concentration of 1.27 $mg.L^{-1}$ in its biomass. It was also suggested that iodate uptake in short-term experiments is insignificant compared to iodide uptake. However, our results clearly indicate that in our long-term experiment, iodate is bioaccumulated to the same extent as iodide (Figure 6).

The evaluation of iodine removal efficiency by various natural materials from aqueous media is relevant for both the decontamination of water bodies and the evaluation of geochemical behaviour of iodide and iodate in waters (Hu et al., 2005; Sarri et al., 2013). The iodine immobilisation efficiency of our selected microscopic filamentous fungi, depicted in Figure 7, was not higher than 52% of the initial 1 $mg.L^{-1}$ iodate or iodide concentration in the culture medium with the highest removal efficiency determined for strain *A. alternata*.

Surprisingly, the significantly higher iodate removal efficiency by fungal biomass contradicts our bioaccumulation results demonstrated in Figure 6. This indicates that the total loss of iodine from the culture medium, as indicated in Figure 7, was much higher than the loss calculated solely from the iodine concentration accumulated in the biomass. This phenomena is most likely because of the extracellular or intracellular transformation of iodate and iodide into volatile species during incubation (Amachi et al., 2003), which is most likely more efficient for iodate. To support this conclusion, the amount of biovolatilised iodine was calculated indirectly based on the iodine content in the cultivation media after cultivation and the amount of iodine accumulated in the fungal biomass. Some of our previous research already showed that microscopic filamentous fungi are able to efficiently transform several toxic and essential elements from the environment into volatile organic forms throughout their metabolic activity. Such transformations can be mediated both inside or outside microbial cells (Boriová et al., 2015; Urík et al., 2014).

In the case of indirect biovolatilisation outside the cells, various metabolic products are released by the fungus into the medium such as several organic acids, which can interact with other substances present in the cultivation media. To date, the biovolatilisation mechanism of iodine by fungi has not yet been sufficiently explained, mostly because of the fact that both intracellular and extracellular transformation can occur. It has also been suggested that acidic conditions are more favorable for iodide volatilisation and oxidation (Evans et al., 1993). However, in all cases, the metabolically active fungus is responsible for iodine volatilisation, because no significant loss of iodine was detected in control experiments.

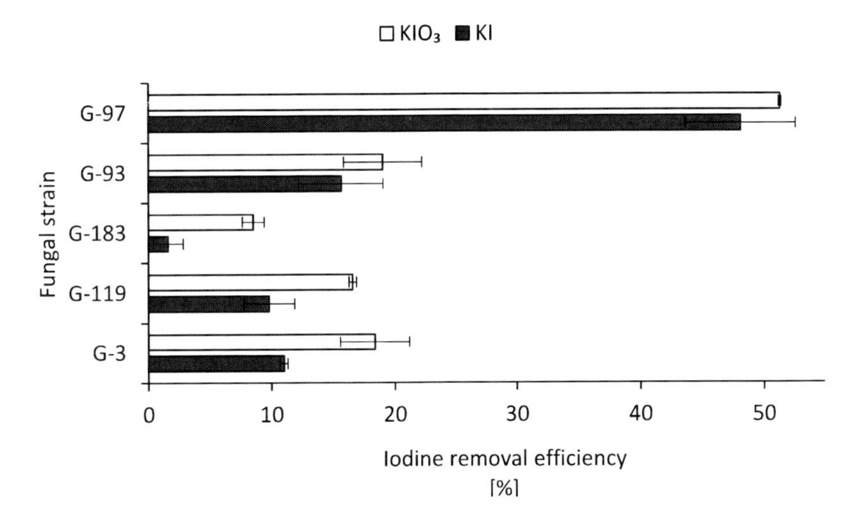

Figure 7. Iodine removal efficiency from the culture media after 25-day cultivation. (G-3 - *Cladosporium cladosporioides*; G-119 - *Aspergillus clavatus*; G-183 - *Penicillium citrinum*; G-93 - *Fusarium oxysporum*; G-97 - *Alternaria alternata*).

The biovolatilisation efficiency of iodine for each fungal strain is presented in Figure 8. The obtained results also indicate that the transformation of iodine into volatile iodine compounds is not only species, but also strain specific. There is also the possibility that not all fungal strains possess volatilisation activity as observed for the fungal strain *P. citrinum*. This can be due to the fungus' different mechanism of iodine metabolism or detoxification (Urík et al., 2007).

The most efficient producers of volatile iodine compounds showed to be strains of *A. alternata* and *F. oxysporum* with average volatilisation of 36.3 and 32 $\mu g.g^{-1}$, respectively, when iodine was added to the medium as iodate. Surprisingly, the average volatilisation rate from iodide treatments was approximately half that amount. A statistically significant difference was observed for the strains of *A. alternata* and *C. cladosporioides*. This effect could be due to the different mechanisms of uptake and metabolic transformation of iodide and iodate by the selected fungi, which is still unclear. Based on our experimental data, we can assume that in the case of iodide, the main detoxification mechanism of the fungi is probably the efflux of the accumulated iodide back to the media, while iodate is preferentially transformed into volatile forms and volatilised into the fungal headspace (Figure 8). In the study by Ban-nai et al. (2006), the strains *C. cladosporioides* and *A. alternata* also volatilised iodine after a 49-day cultivation and the chemical form of volatilised iodine was identified as methyl iodide (CH_3I) while no other possible organic iodine compounds volatilised were detected in their study.

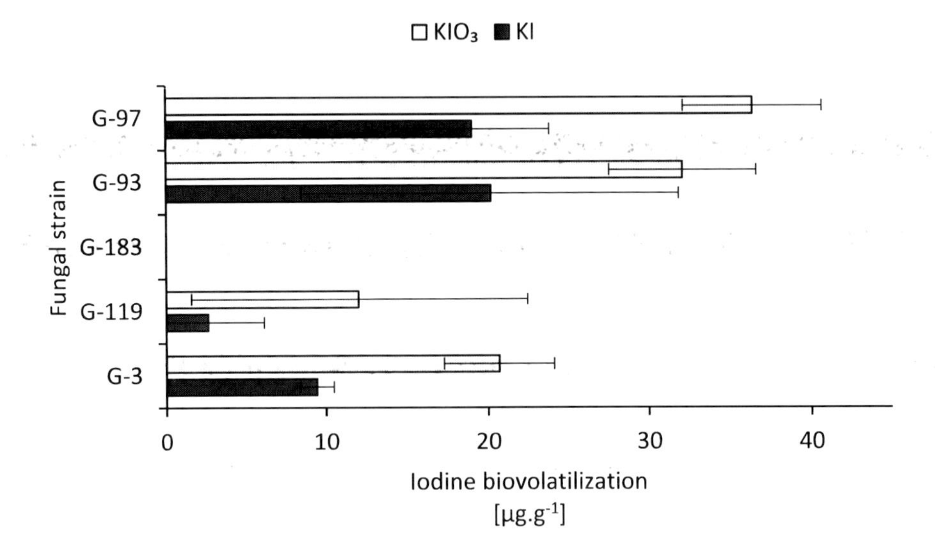

Figure 8. Iodine volatilization efficiency of the selected fungal strains after the 25-day cultivation. (G-3 - *Cladosporium cladosporioides*; G-119 - *Aspergillus clavatus*; G-183 - *Penicillium citrinum*; G-93 - *Fusarium oxysporum*; G-97 - *Alternaria alternata*).

3.4.4. Conclusion

Our results indicate that microscopic filamentous fungi are capable of accumulating iodine regardless of its chemical form. The comparison of the apparent bioaccumulation rate of iodide and iodate for each fungal strain separately after a 25-day cultivation showed that both species are accumulated approximately to the same extent. Our results also highlighted the fungal ability to transform iodine into volatile species. However, not all of our fungal strains showed this ability, thus indicating that iodine volatilisation is strain specific. Unlike bioaccumulation, the iodine volatilisation efficiency was significantly affected by iodine species supplemented in culture medium, therefore, the observed volatilisation rate of iodate was twice as high as iodide.

4. IODINE IN PLANTS

Plants generally contain only trace amounts of iodine, but they still represent an important iodine source for grazing livestock (Lidiard, 1995). Despite iodine's necessity for humans, it was proved to be toxic for plants, however, it can still be beneficial at very low concentrations for plant growth (Kabata-Pendias, 2010; Umaly and Poel, 1970). The positive effect of iodine on plants, however is rare and there is an overwhelming number of scientific papers that have shown that plants are sensitive to iodine and that their exposure even to low iodine concentrations resulted in significantly lower crop yields (Lawson et al., 2015; Mackowiak and Grossl, 1999). Nevertheless, the aim of recent available studies is to seek out suitable crops to produce iodine-fortified plants as an alternative method for iodine supplementation to the human population, in addition to iodinated salt (Caffagni et al., 2011; Caffagni et al., 2012; Comandini et al., 2013; Smoleń and Sady, 2012; Weng et al., 2013).

The recent available data about iodine toxicity in plant species is limited. Tensho and Yeh (1970) were the first to highlight that iodine is toxic for plants. They suggested that iodine is responsible for the "Reclamation Akagare" phenomenon, a disease of lowland rice occurring in some paddy fields in particular areas of Japan. Akagare symptoms are leaf chlorosis and necrotic spots in reddish tint on leaves. These are similar to those symptoms observed in the case of bromine excess in soils or in several other cultivation substrates (Kabata-Pendias, 2010; Watanabe and Tensho, 1970).

Iodine can be taken up by plants via various routes - by root uptake from soil solutions, or by leaves from iodine atmospheric deposition and precipitation (Lawson et al., 2015; Šeda et al., 2012; Whitehead, 1975). Lichens and mosses can accumulate up to 1 $mg.kg^{-1}$ of iodine from aerial fallout (Korobova, 2010) and are thus considered as iodine bioindicators and efficient accumulators. Positive correlation between soil iodine content and plant iodine concentration has also been reported (Sheppard and Motycka, 1997; Weng et al., 2008a). For example, iodine levels in pakchoi, celery, pepper and radish increased with increasing soil iodine concentrations (Hong et al., 2009). However, plants can accumulate iodine from hydroponic cultures and liquid cultivation media much easier than from soils (Zhu et al., 2003). Ashworth et al. (2003) suggested that only about 0.1% of soil iodine is transferred to plant biomass.

It was proved that plant iodine uptake is greater from soils fertilized with the inorganic iodide species compared to more complex fertilizers composed of iodine-rich seaweeds. However, Smolen and Sady (2012) suggested that soil fertilization with iodate is not an effective method for spinach biofortification. Increased iodine uptake in leaf and fruit vegetables was also observed after application of iodized fertilizer made from algae, although the roots can accumulate significantly more iodine than edible plant parts (Weng et al., 2013).

The approach of producing iodine-rich plants via biofortification methods using various iodine containing fertilizers or irrigation water enriched with iodine is very popular and scientifically supported (Caffagni et al., 2011; Caffagni et al., 2012; Comandini et al., 2013; Smoleń and Sady, 2012; Weng et al., 2013). According to recent studies, tomatoes and potatoes are considered to be the most suitable target for iodine biofortification because of their fairly good accumulation abilities and frequent consummation (Caffagni et al., 2011; Kiferle et al., 2013; Landini et al., 2011; Lawson et al., 2015; Weng et al., 2008b).

4.1. Case Study: Iodine Bioaccumulation by Barley

4.1.1. Introduction

The limited information about iodine toxicity and bioaccumulation rate in plant species inspired us to study the toxic effects of iodide and iodate and their transfer to plant tissues of barley (*Hordeum vulgare* L.). We also studied the effects of cultivation substrate type (spiked soil and agar medium) on iodine bioaccumulation during a short-term laboratory experiment. In order to study iodide and iodate toxicity effects on barley, selected growth parameters, photosynthetic pigment quality and iodine bioaccumulation efficiency was evaluated. To evaluate the influence of natural geochemical processes in soil on iodine bioaccumulation efficiency, we also evaluated the effect of aging on iodine transfer from soil to plant in iodine and iodate spiked soils.

4.1.2. Materials and Methods

Grains of barley (*Hordeum vulgare* L., v. spring, cv. Signora) were acquired from Sempol Ltd. (Bratislava, Slovakia).

The soil sample was collected from the upper 5-20 cm of agricultural area and classified as calcaric Chernozem. Before each experiment, the soil was air-dried in the laboratory at 25°C. Prior to soil analysis, the soil was sifted through a 2 mm sieve and the rocks, roots and debris were removed. To determine soil texture, the pipette method was applied (Fiala, 1999). Soil pH value was determined in deionized water and a 1 $mol.L^{-1}$ KCl solution potentiometrically. $CaCO_3$ content was determined using Janko's calcimeter (Fiala, 1999) and total organic carbon content (TOC) was determined according to Walkley and Black (1934). Obtained pedogeochemical characteristics of the soil are listed in Table 3.

For pot and agar experiments (Figure 9), the iodide and iodate stock solutions were prepared by dissolution of KI and KIO_3 (p.a. Centralchem,

Bratislava, Slovakia) in redistilled water. In the pot experiments, 150 g of air-dried 2 mm sieved soil was added to 300 mL plastic pots (height of 10.5 cm, diameter 8 cm) and mixed with 50 mL iodate or iodide solutions in order to obtain soil iodine concentrations of 0, 5, 25, 50, 100, 200, and 300 $mg.kg^{-1}$. The samples were mixed thoroughly and inoculated with barley grains. Each treatment had 3 replicates with 15 grains in each pot.

Table 3. The selected pedogeochemical soil parameters

Texture	pH_{H2O}	pH_{KCl}	TOC [%]	HA [%]	FA [%]	$CaCO_3$ [%]	Total I [$mg.kg^{-1}$]
loamy	7.98	7.45	2.8	0.52	4.38	3.25	2.69±0.03

TOC-total organic carbon content, HA-humic acid content, FA-fulvic acid content.

In the agar experiments, 50 mL culture media consisting of 25 mL Hoagland nutrient solution (Table 4) and 25 mL iodide or iodate solution in redistilled water were prepared in autoclave glass bottles. The final iodine concentrations were 0, 5, 25, 50, 100, 200 and 300 $mg.kg^{-1}$ in solid agar cultivation media. Afterwards 0.7 g of agar type I (HiMedia, Mumbai, India) was added to each treatment. Bottles were closed and sterilized in autoclave at 120°C for 15 minutes. After sterilization the solutions were placed in a cultivation box and poured into 125x125 mm bioassay dishes. The dishes were immediately closed to prevent iodine volatilisation. After cooling down to room temperature, the agar media hardened. Before inoculation, barley grains were soaked in distilled water for 15 min, disinfected in a 0.5% NaClO solution and rinsed with distilled water. The grains were then introduced into bioassay dishes on the agar cultivation media. Each treatment had 3 replicates with 5 grains in each bioassay dish.

The inoculated bioassay dishes and pots were transferred to an incubator (Pol-eko Aparatura, Wodzisław Śląski, Poland) and left for 24 h in the dark at 25°C; then set to a day-light cycle with an 16/8 h photoperiod at temperature of 25/18°C (day/night), 70% humidity and 300 $\mu mol.m^{-2}.s^{-1}$ photosynthetic active radiation. The pots were irrigated once on the third day of incubation with 10 mL of distilled water. After 6-days of growth, the plants were harvested from the cultivation media, divided into shoots and roots, and the fresh mass, shoot and root lengths were immediately measured. The shoots were then dried in the laboratory at room temperature to a constant weight. This experiment (referred to as the 1st cultivation) was repeated after a 3-month aging period of the collected soil under the same conditions (referred to as the 2nd cultivation).

Chlorophyll *a* and chlorophyll *b* contents were determined after shoot extraction in N,N-dimethylformamide (Alfa Aesar, Germany). Absorbance of the extracts was spectrophotometrically measured at 664 and 647 nm wavelengths using UV-Visible spectrophotometer (Evolution 60S, Thermo Fisher Scientific, Madison, USA). Chlorophyllous pigment levels were calculated as follows (Moran, 1982):

$$Chl_a = 12.64\, A_{664} - 2.99\, A_{647} \quad (3)$$

$$Chl_b = -5.6\, A_{664} + 23.26\, A_{647} \quad (4)$$

where Chl*a* - chlorophyll *a*, Chl*b* - chlorophyll *b*, Chl*t* - total chlorophyll contents expressed in $\mu g.ml^{-1}$. The chlorophyll content which referred to fresh biomass was calculated.

The uptake of iodine by plants from soil and agar media were described by the substrate to plant transfer factor (TF), which in this case was calculated as iodine concentration in the shoot biomass of barley ($mg.kg^{-1}$) and divided by the remaining iodine concentration in the substrate ($mg.kg^{-1}$).

Table 4. The chemical composition of Hoagland nutrient solution

Macronutrient	Concentration [mmol.L^{-1}]	Micronutrient	Concentration [μmol.L^{-1}]
$NH_4H_2PO_4$	1	H_3BO_3	46.2
KNO_3	6	$MnCl_2.4H_2O$	9.15
$Ca(NO_3)_2$	4	$ZnSO_4.7H_2O$	0.77
$MgSO_4$	2	$CuSO_4.5H_2O$	0.32
		$H_2MoO_4.H_2O$	0.11
		Fe-EDTA	22.4

Figure 9. Experimental design of short-termed (a) agar and (b) soil cultivation of barley.

For total soil iodine determination, tetramethylammoniumhydroxide (TMAH) (p.a., Alfa Aesar, Germany) extraction modified according to Yamada et al. (1996) was used. 100 mg of finely powdered (< 1 mm) soil was added to 15 mL centrifugation tubes with 5 mL 5% TMAH, placed in a dry bath incubator (MK-20, Hangzhou Allsheng Instruments Co., Ltd., China) for 4 hours at a temperature of 70°C. The samples were shaken manually approximately every 30 min. After 4-hours of incubation, the samples were centrifuged (CM-6MT, Sky Line, ELMI, USA) at 2300 rpm for 30 min. the supernatant was diluted with redistilled water to 50 mL to a final 1% TMAH concentration in the solution.

For total iodine determination in plant tissue, single TMAH extraction was used. The 50 - 210 mg weight (depending on the treatment's crop yield) of dried pulverized barley stem tissue was added to 15 mL centrifugation tubes with 2 mL 25% TMAH for 4 hours in a dry bath at 70°C and shaken every 30 min. After 4 hours, the samples were rinsed out and diluted to 50 mL with redistilled water, centrifuged and passed through a 0.45 μm pore sized cellulose membrane filter. The extraction method for agar cultivation media was the same one used on plant tissue.

To determine the water soluble iodine fraction after aging, desorption experiments were carried out. 2 g of fine powdered soil was shaken for 24 h with 10 mL of redistilled water. After 24 h, the samples were centrifuged at 2300 rpm for 30 min, the supernatant was filtered then diluted with water and TMAH was added to reach 0.5% m/v in the solution.

Table 5. ICP-MS operating conditions for iodine determination

Sample introduction	
Sample uptake	0.88 $mL.min^{-1}$
Nebulizer type	Cross-flow
Nebulizer gas	Argon, 0.9 $L.min^{-1}$
ICP	
RF power	1200 W
Pulse stage voltage	Auto Lens used
Plasma gas	Argon, 15 $L.min^{-1}$
Auxiliary gas	Argon, 1.2 $L.min^{-1}$
Cones	Nickel
Data Acquisition	
Acquisition mode	Peak hopping
Dwell time	50 ms
Integration time	1000 ms
Limit of detection	0.01 $\mu g.L^{-1}$
Detection error	± 1.5%

The total iodine in the soil and plant TMAH extracts was analyzed by ICP-MS instrument (Perkin Elmer Elan 6000, USA) using Te as an internal standard. The accuracy of the methods was evaluated using certified reference materials (CRM). The optimised measurement conditions are shown in Table 5.

4.1.3. Results and Discussion

4.1.3.1. Toxic Effects of Iodide and Iodate on Barley

Figure 10 depicts the adverse effects of iodide and iodate on shoot and root development in agar media and soil pot cultivation. The presented results indicate that no statistically significant difference between the control and iodate amended treatments were observed in shoot length within the selected concentration range. Similarly, the root lengths from the iodate treatments did not significantly differ from the control treatments cultivated in agar media. However, in soil, higher iodate concentrations (above 100 mg.kg^{-1}) had a statistically significant inhibitory effect on root development, although below this concentration level, the iodate added to the soil seemed to be harmless and rather beneficial for root growth.

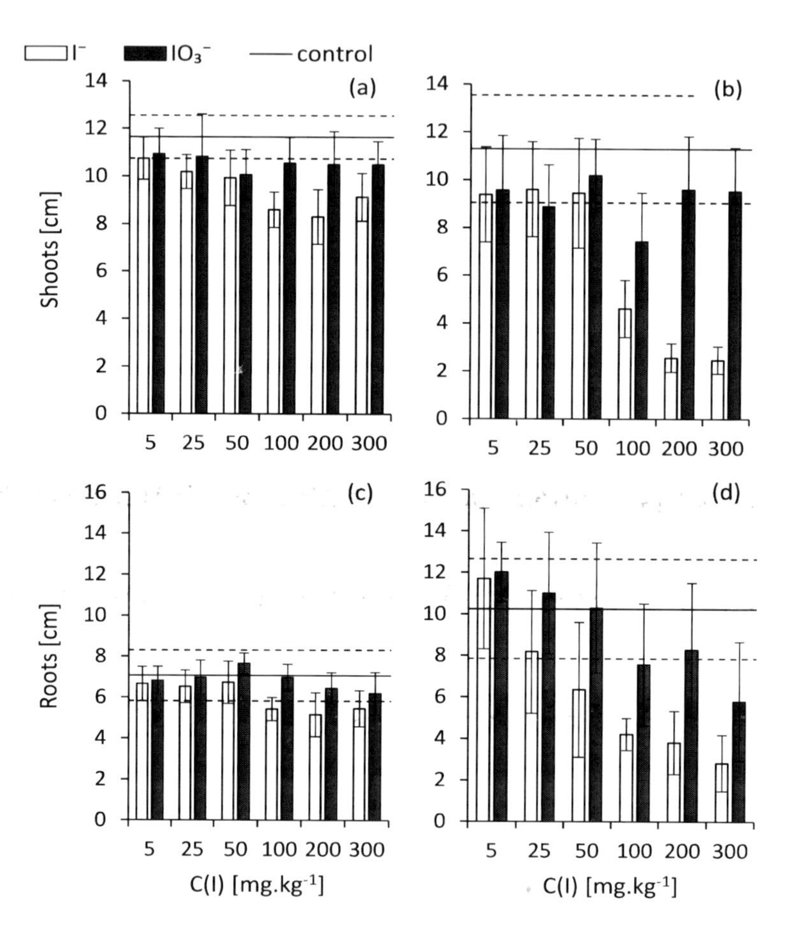

Figure 10. Changes in barley shoot and root growth depending on iodide and iodate concentrations (arithmetic means with standard deviations) in agar cultivation media (a,c) (n = 15) and soil pot cultivation (b,d) (n = 26), * P=0,05, ** P=0,01, *** P=0,001).

Iodide amendment to the system resulted in a significant decrease of shoot length at all concentration levels, and the negative effects of iodide increased with increasing initial iodide concentration. However, iodide effects on root development significantly differ in soil and agar media. While only moderate adverse effects were observed in agar media, the root growth was inhibited in soil at a greater extent.

Our results suggest that the response of barley to the presence of iodine in agar and soil media is more severe for iodide treatments, thus iodide has more distinctive inhibitory effects on plant growth than iodate. This conclusion corresponds appropriately with previous studies on spinach (Dai et al., 2006; Zhu et al., 2003), rice (Blasco et al., 2008; Mackowiak and Grossl, 1999) and lettuce (Blasco et al., 2008). The inhibitory effects of iodate on rice did not occur up to 10 $mg.kg^{-1}$ in cultivation media, however, only 1 $mg.kg^{-1}$ of iodide concentration showed a like hood to cause detrimental effects (Mackowiak and Grossl, 1999).

Lawson et al. (2015) also observed the detrimental effects of iodide on plant growth. They noted a decrease up to 28% in crop yield for butterhead lettuce and kohlrabi when soil was fertilized with iodide. Furthermore, they observed the most common toxicity symptoms for iodine on crops such as chlorosis and necrotic spots on leaves were in the early developmental stages of plants when soil iodide concentration was more than 33 $mg.kg^{-1}$, while iodate was well tolerated. Weng et al. (2008b) observed some adverse effects of iodine on cabbage but only when the soil iodine concentration was more than 50 $mg.kg^{-1}$. These findings suggest that the degree of iodine toxicity is highly dependent on the tolerance of plant species and their uptake mechanism. As it is clear from our experiments, the type of growth substrate may also be a very important factor. In general, iodine uptake is higher from nutrient solutions such as hydroponic systems or agar media. Results presented in this case study indicate that the detrimental effects of both iodine species were substrate-dependent, therefore, the detrimental effects in a soil cultivation system presented to be more obvious. Contrarily Mackowiak et al. (2005) noted that the presence of humic acids in growth solutions can reduce the deleterious effects of iodine on plants by preventing the uptake of high concentrations, since iodine is likely to bind to humic substances which leads to its fixation to humic substances (Dai et al., 2009).

However, both iodide and iodate did not cause a very significant reduction in root length in the agar media; a significant reduction of root hair length was observed even at the lowest initial iodine concentration as shown in Figure 11. The more deleterious effect of iodide compared to iodate on root development could be the result of the fact that iodate is not readily bioavailable for plants and some plant species, such as rice, soybean and also barley, and were proved to be capable of reducing iodate to bioavailable iodide in their rhizosphere (Kato et al., 2013). We assume that iodine uptake from the iodate amended systems is slower due to the biologically driven release of bioavailable forms that are more tolerable for the plant. In addition to biologically mediated iodate reduction, humic substances in soils can also contribute to iodate reduction to molecular iodine and hypoiodous acid. Furthermore, binding of these relatively reactive species onto soil humic substances can also take place (Reiller et al., 2006) which can limit the uptake of iodate by plants.

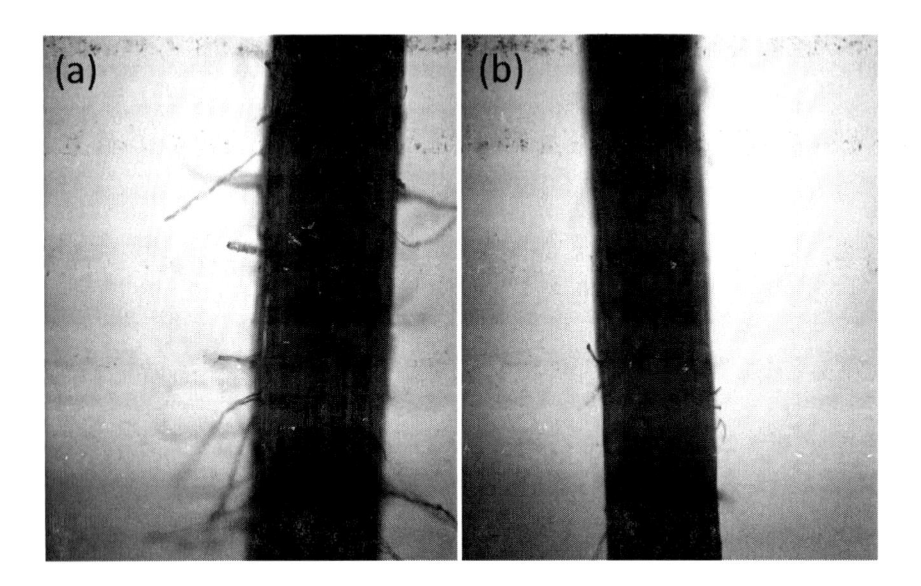

Figure 11. Effects of iodide treatments on barley root hair development from agar cultivation media at an initial iodide concentration of (a) 5 $mg.kg^{-1}$ and (b) .200 $mg.kg^{-1}$.

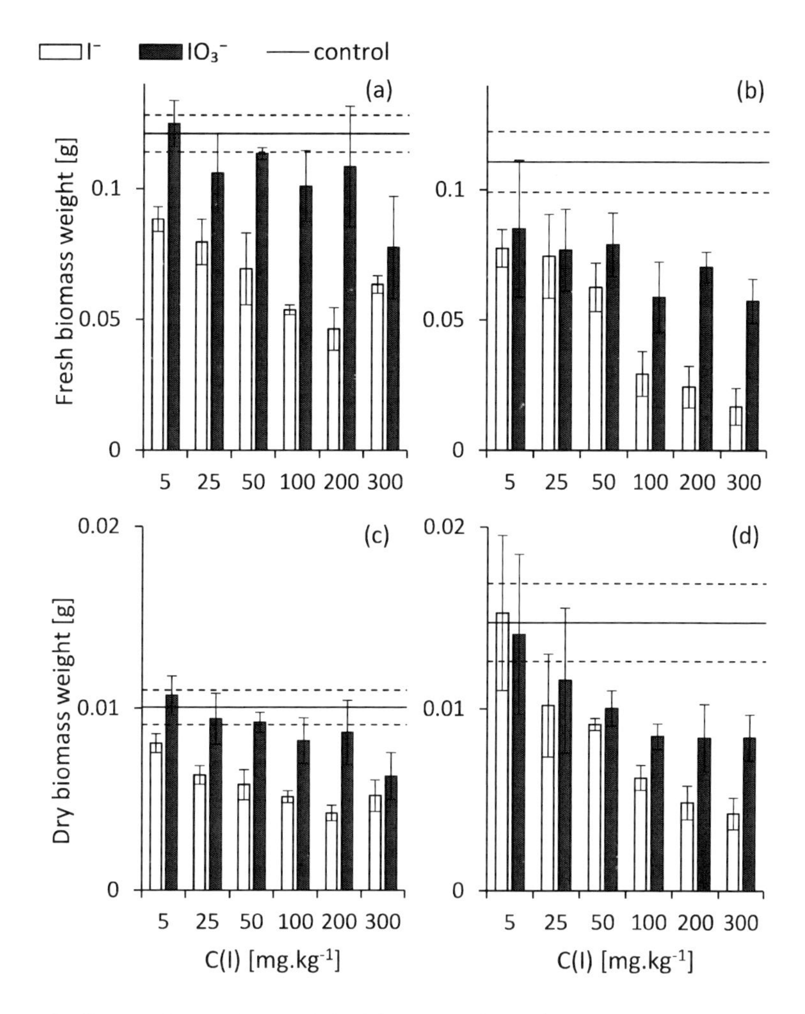

Figure 12. Changes in the fresh and dry biomass weight of barley depending on iodide and iodate concentrations (arithmetic means with standard deviations) in (a,c) agar cultivation media (n = 15) and b) soil pot cultivation (b,d) (n = 26), * P=0,05, ** P=0,01, *** P=0,001).

As highlighted in Figure 12, iodide effects on both fresh and dry biomass synthesis were more severe and statistically significant compared to iodate. In some cases, the biomass was reduced by 85% compared to the control. The response of barley to iodate in the agar media based on its biomass synthesis did not significantly vary from control treatments without iodine in the cultivation media. However, at an iodate concentration of 300 mg.kg^{-1}, a statistically significant biomass synthesis reduction by 48% was observed for both fresh and dry biomass weight. On the contrary, Figure 12b and 12d depict that in soil the detrimental effects of iodate were statistically significant throughout the entire concentration range.

Mild, stimulating effects of both iodate and iodide on barley biomass synthesis at low concentrations indicate that despite its detrimental effects at higher concentrations, it may be beneficial in small concentrations. Despite the deleterious effects of iodides on the biomass production of tomatoes, iodide amendment in cultivation media increased the yield in fruit which led to an increased number of fruits and fruit weight (Lehr et al., 1958). Except for biomass synthesis, iodine amendment can affect some other qualitative characteristics of plants. For example, low concentration of iodide and iodate in fertilizer increased antioxidant response in lettuce (Blasco et al., 2008; Leyva et al., 2011) and was proved to increase ascorbic acid concentration in barbary fig along with an increase of P, Mg, Fe, K, Cu and Mn levels in biomass (Osuna et al., 2014). Iodine amendments also seemed to be beneficial in radish roots, thus free amino acid concentrations in the edible parts increased (Strzetelski et al., 2010), and moreover, caused better accumulation of nitrates(V) and chlorides in spinach, while lowering ammonium and lead accumulation rates (Smoleń and Sady, 2012).

In our experiment, however, iodine enriched agar media did not significantly affect root length, but had distinctive deleterious effects on root hair development (Figure 11). This suggests that there might be a link between the reduction of biomass production and root hair development caused by iodine especially in iodide enriched nutrient solutions since root hairs are considered to play an important role in plant nutrition by facilitating water and nutrient absorption (Minorsky, 2002). Thus, we can assume that root hair reduction inhibits plant water retention. This results in a decrease of fresh biomass weight. In the case of iodate, this effect was not significant and therefore, we suggest that the uptake mechanism of iodide and iodate by barley may be different. The biomass production reduction rate based on dry biomass weight was milder in soil than in agar cultivations. This phenomenon indicates that soil, as a complex matrix, may have a buffer capacity that can be described by the interaction between iodine and soil components, such as the fixation to humic substances or Fe, Mn or Al oxyhydroxides (Dai et al., 2004; Fuhrmann et al., 1998).

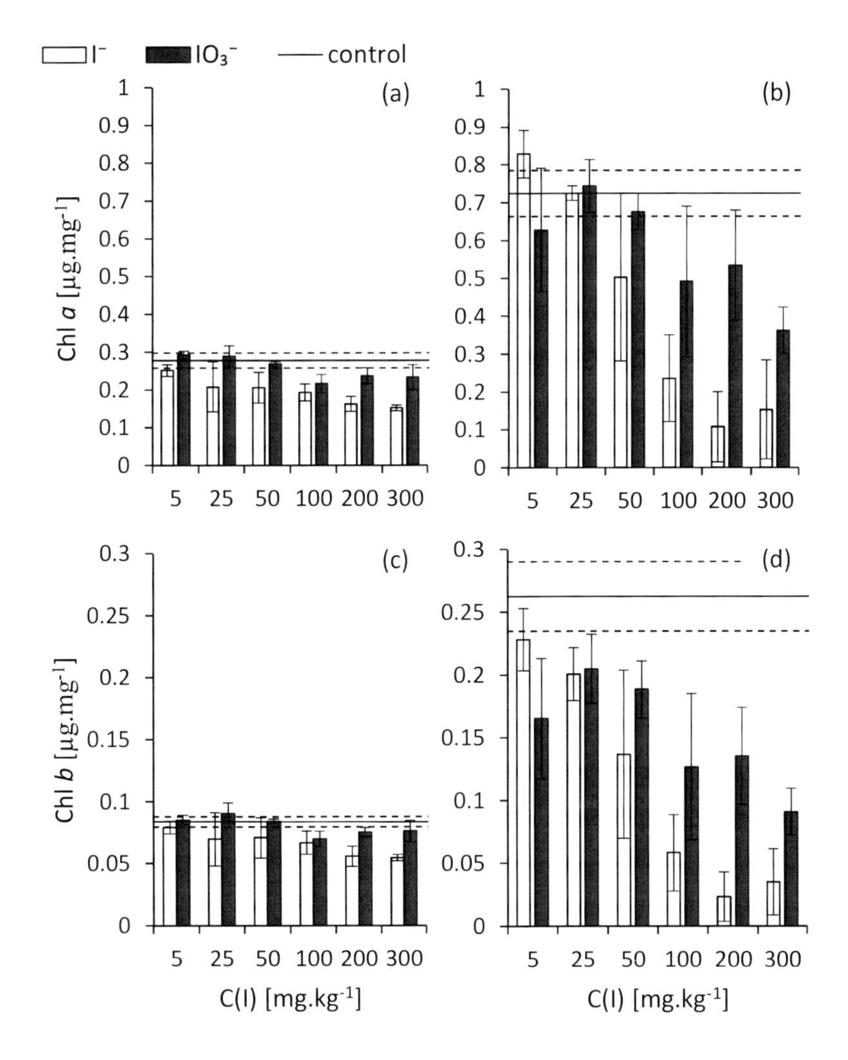

Figure 13. Changes in chlorophyll a and chlorophyll b content depending on iodide and iodate concentrations (arithmetic means with standard deviations) in (a,c) agar cultivation media (n = 5) and (b,d) soil pot cultivation (n = 5), * P=0,05, ** P=0,01, *** P=0,001).

Finally, the changes in chlorophyll *a* and *b* synthesis in agar cultures are highlighted in Figure 13. Likewise as for other growth parameters, iodate had no distinctive effects on chlorophyll *a* synthesis of barley. In soil, chlorophyll *a* inhibitory efficiency was statistically significant only at a concentration level of 200 $mg.kg^{-1}$ and above, where an almost 54% decrease in chlorophyll production was observed compared to the control.

Iodate in agar cultivation media did not cause any significant changes in chlorophyll *a* production by barley up to a concentration level of 50 $mg.kg^{-1}$. Moreover, treatments with the lowest iodate concentration caused a statistically significant increase by 7% in photosynthetic pigment synthesis. In the case of iodide treatments in soil culture, iodide amendment above a concentration level of 50 $mg.kg^{-1}$ caused a 70 to 87% decrease in photosynthetic pigment production. It is important to note that this significant decrease in chlorophyll synthesis was only moderate in agar media compared to the soil system.

Regarding the synthesis of chlorophyll *b*, the trends for both iodide and iodate effects were comparable to the reduction observed chlorophyll *a* synthesis.

A significant chlorophyllous pigment reduction, which is dependent on iodine concentration, was observed in both treatments. As was indicated in some previous studies, iodination of various proteins inside the plant tissue can lead to the inhibition of chlorophyll *a* and chlorophyll *b* production (Mynett and Wain, 1973). Salinity stress can also invoke a decrease in chlorophyllous pigment production (Abdul Qados, 2011). In order to investigate the effect of salinity stress on barley, control experiments were carried out with 1 mM KCl in a nutrient solution, and the results excluded that the inhibition of photosynthetic pigment production developed by salinity stress. Moreover, Leyva et al. (2011) suggested that iodate can be beneficial for lettuce under salinity stress by lowering Na^+ and Cl^- accumulation in biomass, and still, the authors noted that iodate in cultivation substrate caused an increase in biomass production and enhanced the level of soluble sugars in lettuce.

To conclude the observed toxic effects, iodide showed to have more adverse effects on barley than iodate in both agar and soil cultivation systems. This is supported by the IC_{50} values obtained by probit analysis of the observed growth and physiological parameters which are shown in Table 6. These values clearly suggest iodide toxicity in all assessed growth parameters. The result from the calculated data of IC 50 indicate that biomass production and photosynthetic pigment synthesis are more sensitive parameters for the iodine toxicity assessment of barley than for root and shoot length.

Table 6. IC_{50} Values for the assessed growth parameters

	Agar cultivation [mg.kg^{-1}]		**Soil cultivation [mg.kg^{-1}]**	
	I^-	IO_3^-	I^-	IO_3^-
Shoots	2150	NA	87	NA
Roots	6032	4641	94	890
Fresh mass	56	4216	35	655
Dry mass	78	1080	38	860
Chlorophyll (total)	272	2154	85	838

NA – not adjustable within the concentration range.

4.1.3.2. Iodine Bioaccumulation by Barley from Agar Cultivation Media and Soil

Generally, iodide is considered to be more soluble and mobile in the environment (also in soil systems). This means it is more suitable for uptake by plants compared to iodate. In addition, soil sorption capacity for iodate is almost twice as high as for iodide. This should enable efficient plant uptake of iodide (Hong et al., 2009). However, Figure 14 highlights that in the case of the first cultivation, this is only true up to an initial iodide concentration of 100 mg.kg^{-1}. In treatments where the initial iodide concentration was higher, iodate appeared to be up to 3-fold more bioavailable.

The more efficient bioaccumulation by higher iodate concentrations could be the result of two related phenomena: (1) the soil sorption capacity is limited up to 34 mg.kg^{-1} (Dai et al., 2004) allowing for higher iodate concentrations in soil pore water, and (2) iodate reacted with humic substances and soil microorganisms.

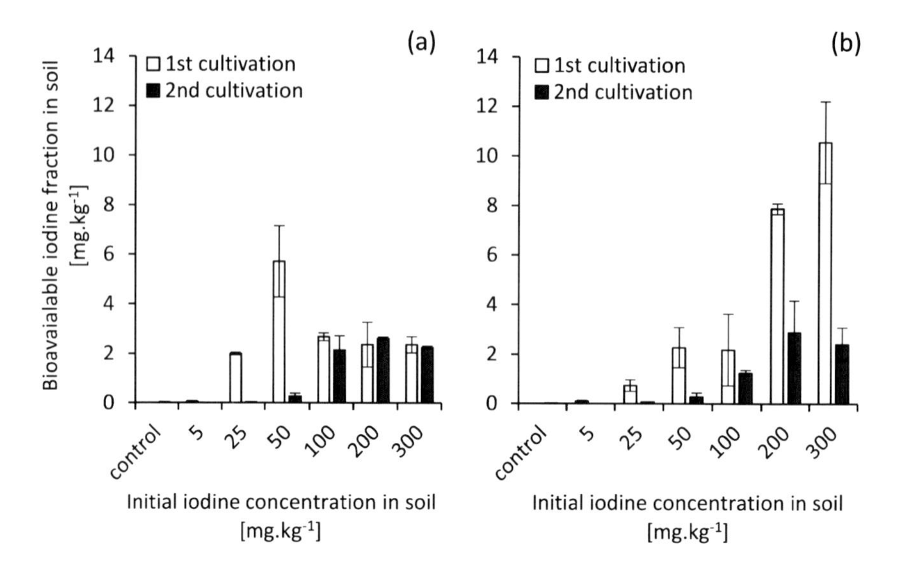

Figure 14. Bioavailable iodine fraction from (a) iodide spiked soil after first and second cultivation (after aging), and (b) iodate spiked soil after first and second cultivation (after aging).

The latter resulted in a (bio)chemical reduction of iodate to molecular iodine and hypoiodous acid, which was proved to be more easily accumulated by plants. Some previous studies suggest that it is necessary to reduce iodate to these forms chemically or biologically in order to promote its uptake by plants (Mackowiak and Grossl, 1999). Conversely, sorption, which can limit the mobility and bioaccumulation rate of these reactive species on humic substances is common in soils with a high organic matter content (Reiller et al., 2006). Moreover, some plant species such as rice and barley have the ability to reduce iodate to iodide in their rhizosphere (Kato et al., 2013). Some microscopic filamentous fungi strains such as *Alternaria* sp. and *Cladosporium* sp. can also accumulate iodine in their tissues and volatilise into the air as methyl iodide which enhances iodine loss from soil (Ban-nai et al., 2006).

Taking into account that the first cultivation was carried out immediately after iodine spiking, the short time between application and cultivation resulted in a limited interaction between iodate and soil particles or microorganisms. Binding of iodine to humic substances, as well as to other soil components, is time demanding (Dai et al., 2004; Dai et al., 2009; Fuhrmann et al., 1998). Furthermore, the absence of the so-called aging process usually leads to significantly higher bioaccumulation rates due to the higher mobility of potentially toxic elements in soils (Hlodák et al., 2016). This is supported by Figure 14b, where after a 3-month aging period, the iodate uptake was reduced practically to the same level as observed in treatments with iodide (Figure 14a).

To highlight the significance of iodine bioavailability and immobilisation in soil, the amount of water soluble fraction was determined by a desorption experiment after aging. As shown in Figure 15, the amount of water soluble iodine increased with increasing total iodine concentration in iodine amended pots. Surprisingly, at lower initial iodine concentrations up to 100 $mg.kg^{-1}$, the iodate treatments' water soluble fractions were slightly higher than in the iodide amended systems. This effect disappeared at higher concentrations, where the water extractable fraction of iodate was significantly lower compared to iodide. This phenomenon reflects the differences in sorption and binding mechanisms of iodide and iodate (Dai et al., 2009). Weak iodine adsorption to soil can result in a higher mobile iodine concentration in the soil solution, which is considered readily available to plants. Iodate, despite its less efficient desorption ability compared to iodide, was still accumulated to the same extent as iodide in plant tissue after aging (Figure 14). This suggests that some other biogeochemical processes are also involved that affect bioavailability and mobility of iodine in soil during aging and cultivation, such as iodine biotransformation into volatile derivatives.

To evaluate the possible differences during iodine's aging process in iodide and iodate fixation potential, the biovolatilisation of iodine from soil during the aging process was also evaluated. The volatilization efficiency of iodine from soil is shown in Figure 16. Iodine volatilisation only occurred from pots where the initial iodine concentrations were higher. A maximum of 22-34% of total iodine was volatilised during aging, however, the difference between the volatilised amounts of iodide and iodate spiked soil is negligible.

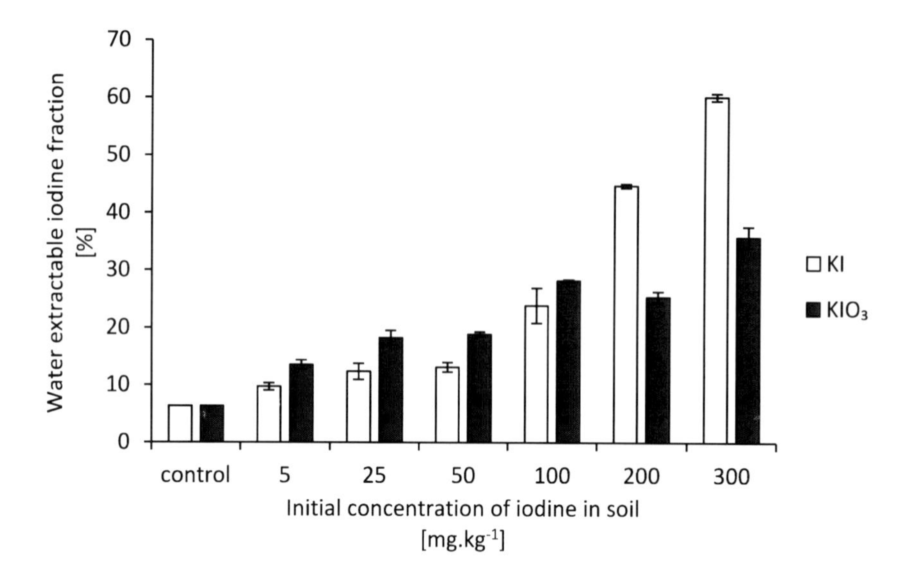

Figurer 15. Relative water soluble iodine fraction in aged soil.

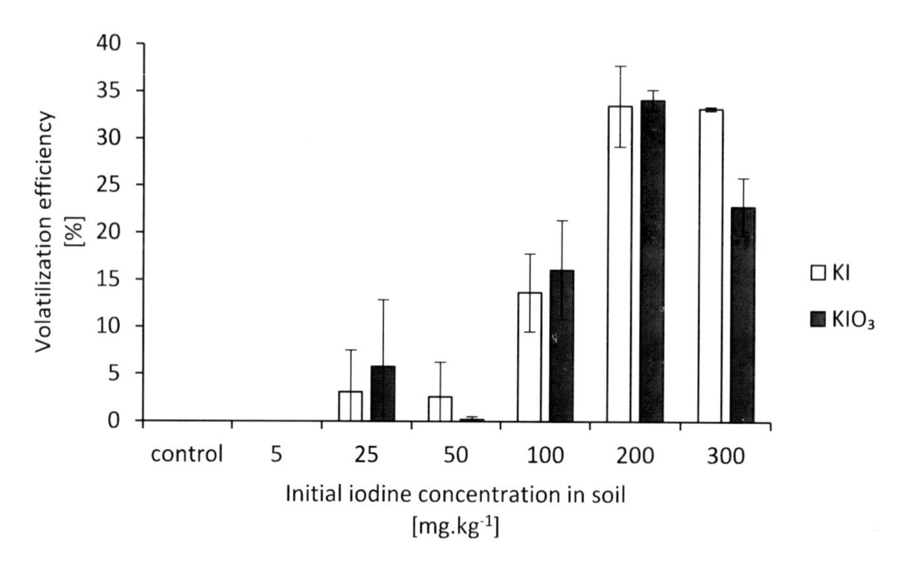

Figure 16. The percentage of volatilised iodine during the aging process in soil.

The Iodine loss through volatilisation from soil can be caused by both the biochemical and physiochemical characteristics of soil. If readily decomposable organic matter is present, iodate may be reduced to elemental iodine, which can then be volatilised into the atmosphere (Redeker et al., 2004b). Microbial activity of soil is another important factor that contributes to high iodine losses and decreased accumulation rates in plants is (Ban-nai et al., 2006).

Generally, the trend line of the obtained transfer factors (TF) corresponds appropriately with the iodine distribution in barley shoots. The TF values were relatively high after the first cultivation, especially in iodide treated samples below 100 $mg.kg^{-1}$ of initial concentration (Figure 17a), where up to 13% of iodine was translocated into plant tissues. This is most likely due to higher mobility of the iodide species (Dai et al., 2009) as well as the short time between spiking and the cultivation. The latter most likely affected the higher iodate bioaccumulation rates from soils during the first cultivation below 100 $mg.kg^{-1}$. However, as it is clear from Figures 17b and 17d, the overall situation has slightly changed during the second cultivation period after aging, where the amount of iodine taken up by plants decreased. Furthermore, no statistical difference was observed between iodide and iodate distribution and TF values by the second cultivation after aging.

However, the mobility of both iodine species significantly differ in agar media. Cultivation in agar media was carried out to evaluate the differences in iodine species bioavailability compared to a heterogeneous soil system, where iodine sorption on soil particles and redox transformations can occur induced by chemical and biological processes (Ban-nai et al., 2006; Fuhrmann et al., 1998). As anticipated, Figure 18 indicates that both evaluated iodine species have a different behaviour in agar media. In agar media, the amount of bioavailable iodine decreased with increasing initial iodine concentration in the media. Overall, the TF values in agar media decreased with increasing initial iodine concentration, however, this trend was not so significant for iodate compared to iodide. Furthermore, iodate bioavailability showed to be lower. In the treatment by 5 $mg.kg^{-1}$ of initial iodide concentration in the cultivation media, approximately 10% of total iodide was accumulated in plant tissue, while in the treatment with only 5 $mg.kg^{-1}$ iodate, only 3% of iodate was translocated to the barley shoots.

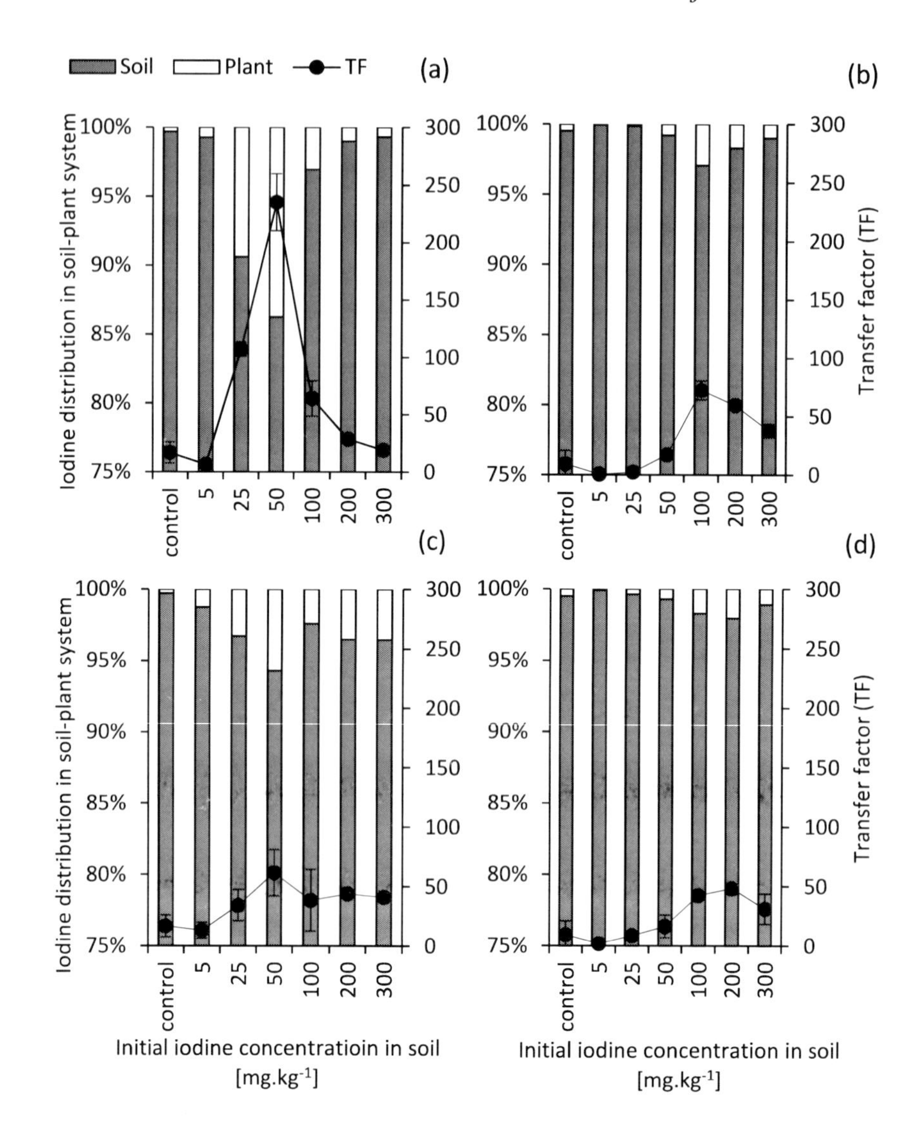

Figure 17. The relative distribution of total iodine in soil-plant system in iodide-spiked (first row) and iodate-spiked (second row) soils after first cultivation (a,c) and after second cultivation (after aging) (b,d) with transfer factor values where similar trend was observed.

Although our TF values for iodate fortified agar cultures are comparable with those obtained by Zhu et al. (2003) for spinach cultivated in hydroponic solution, our TF values for iodide are higher than the reported ones. This highlights both the significance of growth substrate composition and inner physiological response of plants to iodide and iodate in our experiment, which is also reflected in plant growth sensitivity to iodine.

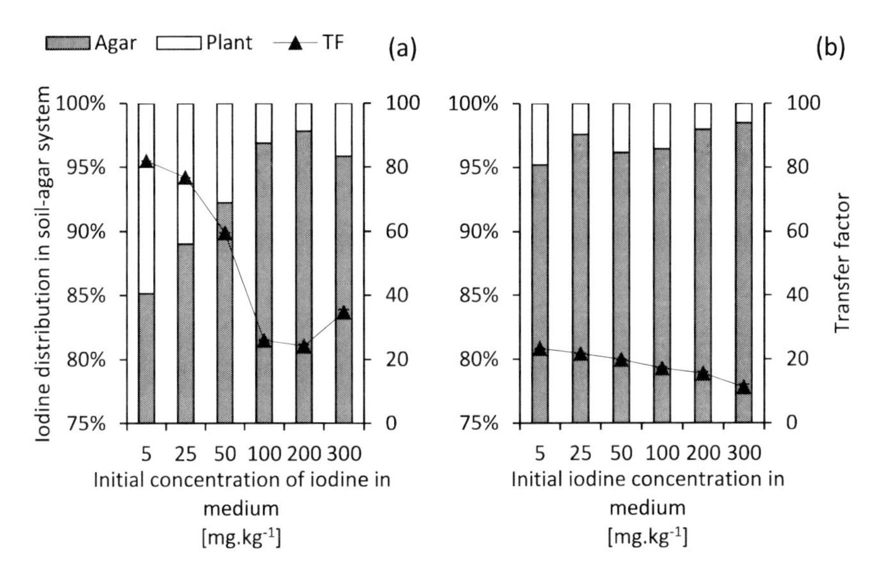

Figure 18. Relative distribution of total iodine in agar-plant system in (a) iodide amended and (b) iodate amended system with transfer factor values.

4.1.4. Conclusion

Our experimental results of iodine toxicity to barley indicate that iodine toxicity is highly dependent on its chemical speciation and substrate type. Generally, both iodine species had adverse effects on every assessed growth parameter, while these effects were more significant in samples where iodine was applied as iodide. Iodate amendment resulted only in moderate inhibition of growth parameters on shoot and root length, biomass production and photosynthetic pigment synthesis. In general, the deleterious effects of both iodine species were more significant in a soil-plant system compared to the agar media.

In our bioaccumulation experiments, a relatively high amount of iodine was accumulated in young barley shoots, when cultivation took place immediately after spiking. The iodide was more bioavailable at lower concentrations, while iodate bioaccumulation was more efficient at higher initial concentrations. However, after a 3-month aging period, the amount of bioavailable iodine was reduced and there was almost no difference between iodide and iodate bioavailability. The high accumulation rates from iodine-fortified soil immediately after spiking is most likely due to the high bioavailability of iodine directly from iodine solutions. Furthermore, in soil through the aging period, relatively intensive iodine volatilisation occurred from pots, especially from treatments with a higher initial iodine concentration.

CONCLUSION

This short contribution indicates that biotic environmental factors significantly affect the mobility and transformation of iodide, as well as iodate. Here, we also provide a short review of abiotic geochemical factors that affect sorption and redox transformations of iodine, even though, it only serves as an introduction to this chapter's complex and experimentally supported discussion on the biogeochemical behaviour and toxic effects of iodide and iodate in soil. However, the data provided are still negligible on a global (bio)geochemical scale. Therefore, we can unequivocally conclude that the highlighted effects of mutual interactions of cultured plants or soil filamentous fungi with iodide and iodate play a significant role only on a local scale. Therefore, global extrapolation of our conclusions and observations is limited. However, the local importance of iodine transfer to agricultural plants and fungal response to iodine is not nominal; and it can be found useful and practical for soil biologists, as well as agricultural chemists, who are willing to find new information on iodine behaviour in a soil-plant system and fungal contribution to its transformation into new, more mobile iodine species.

Acknowledgments

This research was supported by funds obtained from the Scientific Grant Agency of Ministry of Education, Science, Research and Sport of the Slovak Republic and the Slovak Academy of Sciences Nos. VEGA 1/0390/19, VEGA 1/0146/18, VEGA 1/0354/19, VEGA 1/0836/15 and VEGA 1/0164/17.

References

Abdul Qados, A. M. S. (2011). Effect of salt stress on plant growth and metabolism of bean plant *Vicia faba* (L.). *Journal of the Saudi Society of Agricultural Sciences, 10,* 7-15.

Aldahan, A., Englund, E., Possnert, G., Cato, I. & Hou, X. L. (2007). Iodine-129 enrichment in sediment of the Baltic Sea. *Applied Geochemistry, 22,* 637-647.

Almeida, M., Filipe, S., Humanes, M., Maia, M. F., Melo, R., Severino, N., da Silva, J. A. L., Fraústo da Silva, J. J. R. & Wever, R. (2001). Vanadium haloperoxidases from brown algae of the *Laminariaceae* family. *Phytochemistry, 57,* 633-642.

Amachi, S., Kamagata, Y., Kanagawa, T. & Muramatsu, Y. (2001). Bacteria mediate methylation of iodine in marine and terrestrial environments. *Applied and Environmental Microbiology, 67,* 2718-2722.

Amachi, S., Kasahara, M., Hanada, S., Kamagata, Y., Shinoyama, H., Fujii, T. & Muramatsu, Y. (2003). Microbial participation in iodine volatilization from soils. *Environmental Science and Technology, 37,* 3885-3890.

Amachi, S., Kimura, K., Muramatsu, Y., Shinoyama, H. & Fujii, T. (2007). Hydrogen peroxide-dependent uptake of iodine by marine *Flavobacteriaceae* bacterium strain C-21. *Applied and Environmental Microbiology, 73,* 7536-7541.

Amachi, S., Minami, K., Miyasaka, I. & Fukunaga, S. (2010). Ability of anaerobic microorganisms to associate with iodine: ^{125}I tracer experiments using laboratory strains and enriched microbial communities from subsurface formation water. *Chemosphere, 79,* 349-354.

Amachi, S., Mishima, Y., Shinoyama, H., Muramatsu, Y. & Fujii, T. (2005a). Active transport and accumulation of iodide by newly isolated marine bacteria. *Applied and Environmental Microbiology, 71,* 741-745.

Amachi, S., Muramatsu, Y., Akiyama, Y., Miyazaki, K., Yoshiki, S., Hanada, S., Kamagata, Y., Ban-nai, T., Shinoyama, H. & Fujii, T. (2005b). Isolation of iodide-oxidizing bacteria from iodide-rich natural gas brines and seawaters. *Microbial Ecology, 49,* 547-557.

Ashworth, D. J., Shaw, G., Butler, A. P. & Ciciani, L. (2003). Soil transport and plant uptake of radio-iodine from near-surface groundwater. *Journal of Environmental Radioactivity, 70,* 99-114.

Attieh, J. M., Hanson, A. D. & Saini, H. S. (1995). Purification and characterization of a novel methyltransferase responsible for biosynthesis of halomethanes and methanethiol in *Brassica oleracea. Journal of Biological Chemistry, 270,* 9250-9257.

Aulakh, M. S., Wassmann, R. & Rennenberg, H. (2001). Methane emissions from rice fields - quantification, mechanisms, role of management, and mitigation options. *Advances in Agronomy.* Academic Press.

Ball, S. M., Hollingsworth, A. M., Humbles, J., Leblanc, C., Potin, P. & McFiggans, G. (2010). Spectroscopic studies of molecular iodine emitted into the gas phase by seaweed. *Atmospheric Chemistry and Phyics., 10,* 6237-6254.

Ban-nai, T., Muramatsu, Y. & Amachi, S. (2006). Rate of iodine volatilization and accumulation by filamentous fungi through laboratory cultures. *Chemosphere, 65,* 2216-2222.

Bell, N., Hsu, L., Jacob, D. J., Schultz, M. G., Blake, D. R., Butler, J. H., King, D. B., Lobert, J. M. & Maier-Reimer, E. (2002). Methyl iodide:

Atmospheric budget and use as a tracer of marine convection in global models. *Journal of Geophysical Research Atmospheres, 107,* 8-1-8-12.

Blake, N. J., Blake, D. R., Sive, B. C., Chen, T.-Y., Rowland, F. S., Collins, J. E., Sachse, G. W. & Anderson, B. E. (1996). Biomass burning emissions and vertical distribution of atmospheric methyl halides and other reduced carbon gases in the South Atlantic region. *Journal of Geophysical Research: Atmospheres, 101,* 24151-24164.

Blasco, B., Rios, J. J., Cervilla, L. M., Sanchez-Rodrigez, E., Ruiz, J. M. & Romero, L. (2008). Iodine biofortification and antioxidant capacity of lettuce: Potential benefits for cultivation and human health. *Annals of Applied Biology, 152,* 289-299.

Bluhm, K., Croot, P., Wuttig, K. & Lochte, K. (2010). Transformation of iodate to iodide in marine phytoplankton driven by cell senescence. *Aquatic Biology, 11,* 1-15.

Boriová, K., Urík, M., Bujdoš, M. & Matúš, P. (2015). Bismuth(III) volatilization and immobilization by filamentous fungus *Aspergillus clavatus* during aerobic incubation. *Archives of Environmental Contamination and Toxicology, 68,* 405-411.

Brusseau, M. L. & Chorover, J. (2006). Chemical processes affecting contaminant transport and fate. In: Pepper, I. L., Gerba, P. C. & Brusseau, M. L. (Eds.), *Environmental and Pollution Science*, 2nd ed. San Diego: Elsevier Science Academic Press.

Butler, E. C. V., Smith, J. D. & Fisher, N. S. (1981). Influence of phytoplankton on iodine speciation in seawater. *Limnology and Oceanography, 26,* 382-386.

Caffagni, A., Arru, L., Meriggi, P., Milc, J., Perata, P. & Pecchioni, N. (2011). Iodine fortification plant screening process and accumulation in tomato fruits and potato tubers. *Communications in Soil Science and Plant Analysis, 42,* 706-718.

Caffagni, A., Pecchioni, N., Meriggi, P., Bucci, V., Sabatini, E., Acciarri, N., Ciriaci, T., Pulcini, L., Felicioni, N. & Beretta, M. (2012). Iodine uptake and distribution in horticultural and fruit tree species. *Italian Journal of Agronomy, 7,* 32.

Carpenter, L. J., Malin, G., Liss, P. S. & Küpper, F. C. (2000). Novel biogenic iodine-containing trihalomethanes and other short-lived halocarbons in the coastal east Atlantic. *Global Biogeochemical Cycles, 14,* 1191-1204.

Challenger, F. (1951). Biological methylation. *Advances in Enzymology, 12,* 432-491.

Chameides, W. L. & Davis, D. D. (1980). Iodine: Its possible role in tropospheric photochemistry. *Journal of Geophysical Research: Oceans, 85,* 7383-7398.

Christiansen, J. V. & Carlsen, L. (1989). *Iodine in the Environment Revisited. An Evaluation of the Chemical- and Physico Chemical Processes Possibly Controlling the Migration Behaviour of Iodine in the Terrestrial Environment.* Roskilde: Grafisk Service.

Colin, C., Leblanc, C., Wagner, E., Delage, L., Leize-Wagner, E., Van Dorsselaer, A., Kloareg, B. & Potin, P. (2003). The brown algal kelp *Laminaria digitata* features distinct bromoperoxidase and iodoperoxidase activities. *Journal of Biological Chemistry, 278,* 23545-23552.

Comandini, P., Cerretani, L., Rinaldi, M., Cichelli, A. & Chiavaro, E. (2013). Stability of iodine during cooking: investigation on biofortified and not fortified vegetables. *International Journal of Food Sciences and Nutrition, 64,* 857-861.

Councell, T. B., Landa, E. R. & Lovley, D. R. (1997). Microbial reduction of iodate. *Water, Air, and Soil Pollution, 100,* 99-106.

Couture, R. A. & Seitz, M. G. (1983). Sorption of anions of iodine by iron oxides and kaolinite. *Nuclear and Chemical Waste Management, 4,* 301-306.

Čurlík, J. & Jurkovič, L. (2012). *Pedogeochémia,* 1st ed. Univerzita Komenského v Bratislave, Bratislava.

Dai, J. L., Zhang, M. & Zhu, Y. G. (2004). Adsorption and desorption of iodine by various Chinese soils: I. Iodate. *Environment International, 30,* 525-530.

Dai, J. L., Zhang, M., Hu, Q. H., Huang, Y. Z., Wang, R. Q. & Zhu, Y. G. (2009). Adsorption and desorption of iodine by various Chinese soils: II. Iodide and iodate. *Geoderma, 153,* 130-135.

Dai, J. L., Zhu, Y. G., Huang, Y. Z., Zhang, M. & Song, J. L. (2006). Availability of iodide and iodate to spinach (*Spinacia oleracea* L.) in relation to total iodine in soil solution. *Plant and Soil, 289,* 301-308.

Davidson, A. N., Chee-Sanford, J., Lai, H. Y., Ho, C.-h., Klenzendorf, J. B. & Kirisits, M. J. (2011). Characterization of bromate-reducing bacterial isolates and their potential for drinking water treatment. *Water Research, 45,* 6051-6062.

Davis, D., Crawford, J., Liu, S., McKeen, S., Bandy, A., Thornton, D., Rowland, F. & Blake, D. (1996). Potential impact of iodine on tropospheric levels of ozone and other critical oxidants. *Journal of Geophysical Research: Atmospheres, 101,* 2135-2147.

Dimmer, C. H., Simmonds, P. G., Nickless, G. & Bassford, M. R. (2001). Biogenic fluxes of halomethanes from Irish peatland ecosystems. *Atmospheric Environment, 35,* 321-330.

El-Sayed, M. T. (2015). An investigation on tolerance and biosorption potential of *Aspergillus awamori* ZU JQ 965830.1 TO Cd(II). *Annals of Microbiology, 65,* 69-83.

Emerson, H. P., Xu, C., Ho, Y.-F., Zhang, S., Schwehr, K. A., Lilley, M., Kaplan, D. I., Santschi, P. H. & Powell, B. A. (2014). Geochemical controls of iodine uptake and transport in Savannah River Site subsurface sediments. *Applied Geochemistry, 45,* 105-113.

Englund, E., Aldahan, A., Hou, X. L., Petersen, R. & Possnert, G. (2010). Speciation of iodine (^{127}I and ^{129}I) in lake sediments. *Nuclear Instruments and Methods in Physics Research Section B: Beam Interactions with Materials and Atoms, 268,* 1102-1105.

Evans, G. J., Mirbod, S. M. & Jervis, R. E. (1993). Volatilization of iodine species over dilute iodide solutions. *Canadian Journal of Chemical Engineering, 71,* 761-765.

Farhangrazi, Z. S., Sinclair, R., Yamazaki, I. & Powers, L. S. (1992). Haloperoxidase activity of *Phanerochaete chrysosporium* lignin peroxidases H2 and H8. *Biochemistry, 31,* 10763-10768.

Farrenkopf, A. M., Dollhopf, M. E., Chadhain, S. N., Luther, G. W. & Nealson, K. H. (1997). Reduction of iodate in seawater during Arabian Sea shipboard incubations and in laboratory cultures of the marine bacterium *Shewanella putrefaciens* strain MR-4. *Marine Chemistry, 57,* 347-354.

Fiala, K. (1999). *Záväzné metódy rozborov pôd,* 1st ed. Bratislava: Výskumný ústav pôdoznalectva a ochrany pôdy Bratislava. [*Binding methods for soil analysis*].

Fox, P. M., Davis, J. A. & Luther Iii, G. W. (2009). The kinetics of iodide oxidation by the manganese oxide mineral birnessite. *Geochimica et Cosmochimica Acta, 73,* 2850-2861.

Fuge, R. & Johnson, C. C. (1986). The geochemistry of iodine - A review. *Environmental Geochemistry and Health, 8,* 31-54.

Fuhrmann, M., Bajt, S. a. & Schoonen, M. A. A. (1998). Sorption of iodine on minerals investigated by X-ray absorption near edge structure (XANES) and ^{125}I tracer sorption experiments. *Applied Geochemistry, 13,* 127-141.

Furtmüller, P. G., Jantschko, W., Regelsberger, G., Jakopitsch, C., Arnhold, J. & Obinger, C. (2002). Reaction of lactoperoxidase compound I with halides and thiocyanate. *Biochemistry, 41,* 11895-11900.

Fuse, H., Inoue, H., Murakami, K., Takimura, O. & Yamaoka, Y. (2003). Production of free and organic iodine by *Roseovarius* spp. *FEMS Microbiology Letters, 229,* 189-194.

Gallard, H., Allard, S., Nicolau, R., von Gunten, U. & Croué, J. P. (2009). Formation of iodinated organic compounds by oxidation of iodide-containing waters with manganese dioxide. *Environmental Science and Technology, 43,* 7003-7009.

Giles, C. H., Smith, D. & Huitson, A. (1974). A general treatment and classification of the solute adsorption isotherm. I. Theoretical. *Journal of Colloid and Interface Science, 47,* 755-765.

Gozlan, R. S. & Margalith, P. (1973). Iodide oxidation by a marine bacterium. *Journal of Applied Bacteriology, 36,* 407-417.

Greenspan, F. S. (2003). Štítná žláza. In: Greenspan, F. S. & Baxter, J. D. (Eds.), *Základní a klinická endokrinologie*. Praha: H & H. [Thyroid. In: *Basic and Clinical Endocrinology*].

Hansen, V., Roos, P., Aldahan, A., Hou, X. & Possnert, G. (2011). Partition of iodine (^{129}I and ^{127}I) isotopes in soils and marine sediments. *Journal of Environmental Radioactivity, 102,* 1096-1104.

Harper, D. B. (1985). Halomethane from halide ion-a highly efficient fungal conversion of environmental significance. *Nature, 315,* 55-57.

Harper, D. B. & Kennedy, J. T. (1986). Effect of growth conditions on halomethane production by *Phellinus* species: biological and environmental implications. *Microbiology, 132,* 1231-1246.

Hejtmánková, A., Vejdová, M. & Trnková, E. (2005). Stanovení jodu v biologickém materiálu metodou HPLC s elektrochemickým detektorem. [Determination of iodine in biological material by HPLC method with electrochemical detector.] *Chemické Listy*, 99, 657-660.

Hiller, E., Tatarková, V., Šimonovičová, A. & Bartal', M. (2012). Sorption, desorption, and degradation of (4-chloro-2-methylphenoxy) acetic acid in representative soils of the Danubian Lowland, Slovakia. *Chemosphere, 87,* 437-444.

Hlodák, M., Urik, M., Matus, P. & Korenkova, L. (2016). Mercury in mercury(II)-spiked soils is highly susceptible to plant bioaccumulation. *International Journal of Phytoremediation, 18,* 195-199.

Hong, C. l., Weng, H. X., Yan, A. l. & Islam, E. U. (2009). The fate of exogenous iodine in pot soil cultivated with vegetables. *Environmental Geochemistry and Health, 31,* 99-108.

Hong, C., Weng, H., Jilani, G., Yan, A., Liu, H. & Xue, Z. (2012). Evaluation of iodide and iodate for adsorption–desorption characteristics and bioavailability in three types of soil. *Biological Trace Element Research, 146,* 262-271.

Hou, X. (2009). Iodine speciation in foodstuffs, tissues, and environmental samples: Iodine species and analytical method. In: Victor, P., Gerard, B. & Ronald, W. (Eds.), *Comprehensive Handbook of Iodine*. San Diego: Academic Press.

Hu, Q., Zhao, P., Moran, J. E. & Seaman, J. C. (2005). Sorption and transport of iodine species in sediments from the Savannah River and Hanford Sites. *Journal of Contaminant Hydrology, 78,* 185-205.

Hu, Q. H., Moran, J. E. & Blackwood, B. (2009). Geochemical Cycling of Iodine Species in Soils. *Comprehensive Handbook of Iodine: Nutritional, Biochemical, Pathological and Therapeutic Aspects,* 93-105.

Hu, Z. & Moore, R. M. (1996). Kinetics of methyl halide production by reaction of DMSP with halide ion. *Marine Chemistry, 52,* 147-155.

Hughes, C., Franklin, D. J. & Malin, G. (2011). Iodomethane production by two important marine cyanobacteria: *Prochlorococcus marinus* (CCMP 2389) and *Synechococcus* sp. (CCMP 2370). *Marine Chemistry, 125,* 19-25.

Itoh, N., Tsujita, M., Ando, T., Hisatomi, G. & Higashi, T. (1997). Formation and emission of monohalomethanes from marine algae. *Phytochemistry, 45,* 67-73.

Jia-Zhong, Z. & Whitfield, M. (1986). Kinetics of inorganic redox reactions in seawater: I. The reduction of iodate by bisulphide. *Marine Chemistry, 19,* 121-137.

Johnson, C. C. (2003). *The geochemistry of iodine and its application to environmental strategies for reducing the risks from iodine deficiency disorders (IDD). (CR/03/057N).* Nottingham: British Geological Survey.

Kabata-Pendias, A. (2010). Elements of group 17 (Previously Group VIIa). *Trace Elements in Soils and Plants, Fourth Edition.* CRC Press.

Kaplan, D. I., Serne, R. J., Parker, K. E. & Kutnyakov, I. V. (1999). Iodide sorption to subsurface sediments and illitic minerals. *Environmental Science and Technology, 34,* 399-405.

Kato, S., Wachi, T., Yoshihira, K., Nakagawa, T., Ishikawa, A., Takagi, D., Tezuka, A., Yoshida, H., Yoshida, S., Sekimoto, H. & Takahashi, M. (2013). Rice (*Oryza sativa* L.) roots have iodate reduction activity in response to iodine. *Frontiers in Plant Science, 4.*

Kengen, S. W. M., Rikken, G. B., Hagen, W. R., Van Ginkel, C. G. & Stams, A. J. M. (1999). Purification and characterization of (per)chlorate reductase from the chlorate-respiring strain GR-1. *Journal of Bacteriology, 181,* 6706-6711.

Khun, M. & Čerňanský, S. (2011). *Geofaktory a zdravotné aspekty kvality života* [*Geofactors and health aspects of quality of life*], 1st ed. Bratislava: Univerzita Komenského v Bratislave.

Kiferle, C., Gonzali, S., Holwerda, H. T., Ibaceta, R. R. & Perata, P. (2013). Tomato fruits: A good target for iodine biofortification. *Frontiers in Plant Science, 4.*

Korobova, E. (2010). Soil and landscape geochemical factors which contribute to iodine spatial distribution in the main environmental components and food chain in the central Russian plain. *Journal of Geochemical Exploration, 107,* 180-192.

Küpper, F. C., Carpenter, L. J., Leblanc, C., Toyama, C., Uchida, Y., Maskrey, B. H., Robinson, J., Verhaeghe, E. F., Malin, G., Luther, G. W., Kroneck, P. M. H., Kloareg, B., Meyer-Klaucke, W., Muramatsu, Y., Megson, I. L., Potin, P. & Feiters, M. C. (2013). *In vivo* speciation studies and antioxidant properties of bromine in *Laminaria digitata* reinforce the significance of iodine accumulation for kelps. *Journal of Experimental Botany, 64,* 2653-2664.

Küpper, F. C., Carpenter, L. J., McFiggans, G. B., Palmer, C. J., Waite, T. J., Boneberg, E.-M., Woitsch, S., Weiller, M., Abela, R., Grolimund, D., Potin, P., Butler, A., Luther, G. W., Kroneck, P. M. H., Meyer-Klaucke, W. & Feiters, M. C. (2008). Iodide accumulation provides kelp with an inorganic antioxidant impacting atmospheric chemistry. *Proceedings of the National Academy of Sciences of the United States of America, 105,* 6954-6958.

Küpper, F. C., Schweigert, N., Ar Gall, E., Legendre, J. M., Vilter, H. & Kloareg, B. (1998). Iodine uptake in Laminariales involves extracellular, haloperoxidase-mediated oxidation of iodide. *Planta, 207,* 163-171.

Lai, C. Y., Lv, P. L., Dong, Q. Y., Yeo, S. L., Rittmann, B. E. & Zhao, H. P. (2018). Bromate and nitrate bioreduction coupled with poly-beta-hydroxybutyrate production in a methane-based membrane biofilm reactor. *Environemntal Science and Technology, 52,* 7024-7031.

Landini, M., Gonzali, S. & Perata, P. (2011). Iodine biofortification in tomato. *Journal of Plant Nutrition and Soil Science, 174,* 480-486.

Laturnus, F., Adams Freddy, C. & Wiencke, C. (1998). Methyl halides from Antarctic macroalgae. *Geophysical Research Letters, 25,* 773-776.

Laturnus, F., Svensson, T., Wiencke, C. & Öberg, G. (2004). Ultraviolet radiation affects emission of ozone-depleting substances by marine macroalgae: Results from a laboratory incubation study. *Environmental Science & Technology, 38,* 6605-6609.

Lawson, P. G., Daum, D., Czauderna, R., Meuser, H. & Hartling, J. W. (2015). Soil versus foliar iodine fertilization as a biofortification strategy for field-grown vegetables. *Frontiers in Plant Science, 6.*

Lehr, J. J., Wybenga, J. M. & Rosanow, M. (1958). Iodine as a micronutrient for tomatoes. *Plant Physiology, 33,* 421-427.

Leyva, R., Sánchez-Rodríguez, E., Ríos, J. J., Rubio-Wilhelmi, M. M., Romero, L., Ruiz, J. M. & Blasco, B. (2011). Beneficial effects of exogenous iodine in lettuce plants subjected to salinity stress. *Plant Science, 181,* 195-202.

Lidiard, H. M. (1995). Iodine in the reclaimed upland soils of a farm in the Exmoor National-Park, Devon, UK and its inpact on livestock helath. *Applied Geochemistry, 10,* 85-95.

López-Gutiérrez, J. M., Garcia-León, M., Schnabel, C., Suter, M., Synal, H. A., Szidat, S. & Garcia-Tenorio, R. (2004). Relative influence of ^{129}I sources in a sediment core from the Kattegat area. *Science of The Total Environment, 323,* 195-210.

Lusa, M., Bomberg, M., Aromaa, H., Knuutinen, J. & Lehto, J. (2015). Sorption of radioiodide in an acidic, nutrient-poor boreal bog: Insights into the microbial impact. *Journal of Environmental Radioactivity, 143,* 110-122.

Mackowiak, C. L. & Grossl, P. R. (1999). Iodate and iodide effects on iodine uptake and partitioning in rice (*Oryza sativa* L.) grown in solution culture. *Plant and Soil, 212,* 133-141.

Mackowiak, C. L., Grossl, P. R. & Cook, K. L. (2005). Iodine toxicity in a plant-solution system with and without humic acid. *Plant and Soil, 269,* 141-150.

MacLean, L. C. W., Martinez, R. E. & Fowle, D. A. (2004). Experimental studies of bacteria–iodide adsorption interactions. *Chemical Geology, 212,* 229-238.

Maguire, R. J. & Dunford, H. B. (1972). Kinetics of the oxidation of iodide ion by lactoperoxidase compound II. *Biochemistry, 11,* 937-941.

Manavathu, E. K., Dimmock, J. R., Vashishtha, S. C. & Chandrasekar, P. H. (1999). Proton-pumping-ATPase-targeted antifungal activity of a novel conjugated styryl ketone. *Antimicrobial Agents and Chemotherapy, 43,* 2950-2959.

Manley, S. L. & Cuesta, J. L. (1997). Methyl iodide production from marine phytoplankton cultures. *Limnology and Oceanography, 42,* 142-147.

Manley, S. L. & Dastoor, M. N. (1987). Methyl halide (CH_3X) production from the giant kelp, *Macrocystis*, and estimates of global CH_3X production by kelp. *Limnology and Oceanography, 32,* 709-715.

Maxon, H. R., Thomas, S. R. & Samaratunga, R. C. (1997). Dosimetric considerations in the radioiodine treatment of macrometastases and micrometastases from differentiated thyroid cancer. *Thyroid, 7,* 183-187.

Minorsky, P. V. (2002). Mutant studies on root aair function. *Plant Physiology, 129,* 438-439.

Mok, J. K., Toporek, Y. J., Shin, H.-D., Lee, B. D., Lee, M. H. & DiChristina, T. J. (2018). Iodate reduction by *Shewanella oneidensis* does not involve nitrate reductase. *Geomicrobiology Journal, 35,* 570-579.

Moore, M. R. (1994). Photochemical production of methyl iodide in seawater. *Journal of Geophysical Research: Atmospheres, 99,* 16415-16420.

Moore, R. M., Webb, M., Tokarczyk, R. & Wever, R. (1996). Bromoperoxidase and iodoperoxidase enzymes and production of halogenated methanes in marine diatom cultures. *Journal of Geophysical Research C: Oceans, 101,* 20899-20908.

Moran, R. (1982). Formulae for determination of chlorophyllous pigments extracted with n,n-dimethylformamide. *Plant Physiol, 69,* 1376-1381.

Muramatsu, Y. & Hans Wedepohl, K. (1998). The distribution of iodine in the earth's crust. *Chemical Geology, 147,* 201-216.

Muramatsu, Y. & Yoshida, S. (1995). Volatilization of methyl iodide from the soil-plant system. *Atmospheric Environment, 29,* 21-25.

Muramatsu, Y., Yoshida, S., Fehn, U., Amachi, S. & Ohmomo, Y. (2004). Studies with natural and anthropogenic iodine isotopes: iodine distribution and cycling in the global environment. *Journal of Environmental Radioactivity, 74,* 221-232.

Mynett, A. & Wain, R. L. (1973). Herbicidal Aaction of Iodide: Effect on Chlorophyll Content and Photosynthesis in Dwarf Bean Phaseolus Vulgaris. *Weed Research, 13,* 101-109.

Nelson, P. E., Marasas, W. F. O. & Toussoun, T. A. (1983). *Fusarium species: an illustrated manual for identification.* University Park: Pennsylvania State University Press.

Nightingale, P. D., Malin, G. & Liss, P. S. (1995). Production of chloroform and other low molecular-weight halocarbons by some species of macroalgae. *Limnology and Oceanography, 40,* 680-689.

Osuna, H. T. G., Mendoza, A. B., Morales, C. R., Rubio, E. M., Star, J. V. & Ruvalcaba, R. M. (2014). Iodine application increased ascrobic acid content and modified the vascular tissue in *Opuntia ficus-indica* L. *Pakistan Journal of Botany, 46,* 127-134.

Otosaka, S., Schwehr, K. A., Kaplan, D. I., Roberts, K. A., Zhang, S., Xu, C., Li, H.-P., Ho, Y.-F., Brinkmeyer, R., Yeager, C. M. & Santschi, P. H. (2011). Factors controlling mobility of 127I and 129I species in an acidic groundwater plume at the Savannah River Site. *Science of the Total Environment, 409,* 3857-3865.

Palmer, C. J., Anders, T. L., Carpenter, L. J., Küpper, F. C. & McFiggans, G. B. (2005). Iodine and aalocarbon response of *Laminaria digitata* to oxidative stress and links to atmospheric new particle production. *Environmental Chemistry, 2,* 282-290.

Pitt, J. I. & Hocking, A. D. (2009). *Fungi and food spoilage*, 3rd ed. New York: Springer.

Podoba, J. (1962). *Endemická struma na Slovensku*. Bratislava: Veda. [*Endemic struma in Slovakia*].

Raja, M. E. & Babcock, K. L. (1961). On The Soil Chemistry of Radio-iodine. *Soil Science, 91,* 1-5.

Redeker, K. R. & Cicerone, R. J. (2004). Environmental controls over methyl halide emissions from rice paddies. *Global Biogeochemical Cycles, 18.*

Redeker, K. R., Manley, S. L., Walser, M. & Cicerone, R. J. (2004a). Physiological and biochemical controls over methyl halide emissions from rice plants. *Global Biogeochemical Cycles, 18,* GB1007 1001-1014.

Redeker, K. R., Treseder, K. K. & Allen, M. F. (2004b). Ectomycorrhizal fungi: A new source of atmospheric methyl halides? *Global Change Biology, 10,* 1009-1016.

Redeker, K. R., Wang, N., Low, J. C., McMillan, A., Tyler, S. C. & Cicerone, R. J. (2000). Emissions of methyl halides and methane from rice paddies. *Science, 290,* 966-969.

Reiller, P., Mercier-Bion, F., Gimenez, N., Barré, N. & Miserque, F. (2006). Iodination of humic acid samples from different origins. *Radiochimica Acta, 94,* 739.

Ren, Q., Fan, J., Zhang, Z., Zheng, X. & Delong, G. R. (2008). An environmental approach to correcting iodine deficiency: supplementing iodine in soil by iodination of irrigation water in remote areas. *Journal of Trace Elements in Medicine and Biology, 22,* 1-8.

Saini, H. S., Attieh, J. M. & Hanson, A. D. (1995). Biosynthesis of halomethanes and methanethiol by higher plants via a novel methyltransferase reaction. *Plant, Cell & Environment, 18,* 1027-1033.

Samson, R. A. & Frisvad, J. C. (2004). *Penicillium Subgenus Penicillium: New Taxonomic Schemes and Mycotoxins and Other Extrolites*, 49 ed.: Centraalbureau voor Schimmelcultures.

Sarri, S., Misaelides, P., Noli, F., Papadopoulou, L. & Zamboulis, D. (2013). Removal of iodide from aqueous solutions by polyethylenimine- epichlorohydrin resins. *Journal of Radioanalytical and Nuclear Chemistry, 298,* 399-403.

Seki, M., Oikawa, J. I., Taguchi, T., Ohnuki, T., Muramatsu, Y., Sakamoto, K. & Amachi, S. (2013). Laccase-catalyzed oxidation of iodide and formation of organically bound iodine in soils. *Environmental Science & Technology, 47,* 390-397.

Shah, M. M. & Aust, S. D. (1993a). Iodide as the mediator for the reductive reactions of peroxidases. *Journal of Biological Chemistry, 268,* 8503-8506.

Shah, M. M. & Aust, S. D. (1993b). Oxidation of halides by peroxidases and their subsequent reductions. *Archives of Biochemistry and Biophysics, 300,* 253-257.

Sheppard, M. I., Hawkins, J. L. & Smith, P. A. (1996). Linearity of iodine sorption and sorption capacities for seven soils. *Journal of Environmental Quality, 25,* 1261-1267.

Sheppard, M. I. & Thibault, D. H. (1992). Chemical behaviour of iodine in organic and mineral soils. *Applied Geochemistry, 7,* 265-272.

Sheppard, S. C. (2003). Interpolation of solid/liquid partition coefficients, Kd, for iodine in soils. *Journal of Environmental Radioactivity, 70,* 21-27.

Sheppard, S. C. & Motycka, M. (1997). Is the akagare phenomenon important to iodine uptake by wild rice (*Zizania aquatica*)? *Journal of Environmental Radioactivity, 37,* 339-353.

Shimamoto, Y. S., Itai, T. & Takahashi, Y. (2010). Soil column experiments for iodate and iodide using K-edge XANES and HPLC–ICP-MS. *Journal of Geochemical Exploration, 107,* 117-123.

Shimura, H., Haraguchi, K., Miyazaki, A., Endo, T. & Onaya, T. (1997). Iodide uptake and experimental ^{131}I therapy in transplanted undifferentiated thyroid cancer cells expressing the Na+/I− symporter gene. *Endocrinology, 138,* 4493-4496.

Shinonaga, T., Gerzabek, M. H., Strebl, F. & Muramatsu, Y. (2001). Transfer of iodine from soil to cereal grains in agricultural areas of Austria. *Science of the Total Environment, 267,* 33-40.

Smoleń, S. & Sady, W. (2012). Influence of iodine form and application method on the effectiveness of iodine biofortification, nitrogen metabolism as well as the content of mineral nutrients and heavy metals in spinach plants (*Spinacia oleracea* L.). *Scientia Horticulturae, 143,* 176-183.

Smythe-Wright, D., Boswell, S. M., Breithaupt, P., Davidson, R. D., Dimmer, C. H. & Eiras Diaz, L. B. (2006). Methyl iodide production in the ocean: Implications for climate change. *Global Biogeochemical Cycles, 20.*

Solomon, S., Garcia, R. R. & Ravishankara, A. (1994). On the role of iodine in ozone depletion. *Journal of Geophysical Research: Atmospheres, 99,* 20491-20499.

St-Germain, G. & Summerbell, R. (1996). *Identifying Filamentous Fungi: A Clinical Laboratory Handbook.* Redwood City, California: Star Publishing Company.

Stemmler, I., Hense, I., Quack, B. & Maier-Reimer, E. (2014). Methyl iodide production in the open ocean. *Biogeosciences, 11,* 4459-4476.

Strzetelski, P., Smoleń, S., Rozek, S. & Sady, W. (2010). The effect of diverse iodine fertilization on nitrate accumulation and content of selected compounds in radish plants (*Raphanus sativus* L.). *Acta Scientiarum Polonorum. Hortorum Cultus, 2.*

Šeda, M., Švehla, J., Trávníček, J., Kroupová, V., Konečný, R., Fiala, K., Svozilová, M. & Krhovjaková, J. (2012). The effect of volcanic activity of the Eyjafjallajökul volcano on iodine concentration in precipitation in the Czech Republic. *Chemie der Erde - Geochemistry, 72,* 279-281.

Taurog, A., Howells, E. M. & Nachimson, H. I. (1966). Conversion of iodate to iodide *in vivo* and *in vitro*. *Journal of Biological Chemistry, 241,* 4686-4693.

Taurog, A., Chaikoff, I. & Feller, D. (1947). The mechanism of iodine concentration by the thyroid gland: its non-organic iodine-binding capacity in the normal and propylthiouraciltreatedrat. *Journal of Biological Chemistry, 171,* 189-201.

Tensho, K. & Yeh, K.-L. (1970). Radio-iodine uptake by plant from soil with special reference to lowland rice. *Soil Science and Plant Nutrition, 16,* 30-37.

Thayer, J. S. (2002). Biological methylation of less-studied elements. *Applied Organometallic Chemistry, 16,* 677-691.

Tian, R. C., Marty, J. C., Nicolas, E., Chiavérini, J., Ruiz-Ping, D. & Pizay, M. D. (1996). Iodine speciation: a potential indicator to evaluate new production versus regenerated production. *Deep Sea Research Part I: Oceanographic Research Papers, 43,* 723-738.

Toyohara, M., Kaneko, M., Mitsutsuka, N., Fujihara, H., Saito, N. & Murase, T. (2002). Contribution to understanding iodine sorption mechanism onto mixed solid alumina cement and calcium compounds. *Journal of Nuclear Science and Technology, 39,* 950-956.

Truesdale, V. W. (1978). Iodine in inshore and off-shore marine waters. *Marine Chemistry, 6,* 1-13.

Truesdale, V. W. & Bailey, G. W. (2002). Iodine distribution in the Southern Benguela system during an upwelling episode. *Continental Shelf Research, 22,* 39-49.

Truesdale, V. W. & Luther, G. W. (1995). Molecular iodine reduction by natural and model organic substances in seawater. *Aquatic Geochemistry, 1,* 89-104.

Tsunogai, S. & Sase, T. (1969). Formation of iodide-iodine in the ocean. *Deep Sea Research and Oceanographic Abstracts, 16,* 489-496.

Um, W., Serne, R. J. & Krupka, K. M. (2004). Linearity and reversibility of iodide adsorption on sediments from Hanford, Washington under water saturated conditions. *Water Research, 38,* 2009-2016.

Umaly, R. C. & Poel, L. W. (1970). Effects of various concentrations of iodine as potassium iodide on the growth of barley, tomato and pea in nutrient solution culture. *Annals of Botany, 34,* 919-926.

Urík, M., Boriová, K., Bujdoš, M. & Matúš, P. (2016). Fungal selenium(VI) accumulation and biotransformation - Filamentous fungi in selenate contaminated aqueous media remediation. *Clean - Soil, Air, Water, 44,* 610-614.

Urík, M., Čerňanský, S., Ševc, J., Šimonovičová, A. & Littera, P. (2007). Biovolatilization of arsenic by different fungal strains. *Water, Air, and Soil Pollution, 186,* 337-342.

Urík, M., Hlodák, M., Mikušová, P. & Matúš, P. (2014). Potential of microscopic fungi isolated from mercury contaminated soils to accumulate and volatilize mercury(II). *Water, Air, & Soil Pollution, 225,* 1-11.

Van Bergeijk, S. A., Hernández Javier, L., Heyland, A., Manchado, M. & Pedro Cañavate, J. (2013). Uptake of iodide in the marine haptophyte *Isochrysis* sp. (T.ISO) driven by iodide oxidation. *Journal of Phycology, 49,* 640-647.

Venturi, S., Donati, F. M., Venturi, A. & Venturi, M. (2000). Environmental iodine deficiency: A challenge to the evolution of terrestrial life? *Thyroid, 10,* 727-729.

Venturi, S. & Venturi, M. (2009). Iodine in Evolution of Salivary Glands and in Oral Health. *Nutrition and Health, 20,* 119-134.

Walkley, A. & Black, I. A. (1934). An examination of the Degtjareff method for determining soil organic matter, and a proposed modification of the chromic acid titration method. *Soil Science, 37,* 29-38.

Watanabe, I. & Tensho, K. (1970). Further study on iodine toxicity in relation to "Reclamation Akagare" disease of lowland rice. *Soil Science and Plant Nutrition, 16,* 192-194.

Weng, H. X., Liu, H. P., Li, D. W., Ye, M., Pan, L. & Xia, T. H. (2014). An innovative approach for iodine supplementation using iodine-rich phytogenic food. *Environmental Geochemistry and Health, 36,* 815-828.

Weng, H. X., Yan, A. L., Hong, C. L., Qin, Y. C., Pan, L. & Xie, L. L. (2008a). Biogeochemical transfer and dynamics of iodine in a soil–plant system. *Environmental Geochemistry and Health, 31,* 401-411.

Weng, H., Hong, C., Xia, T., Bao, L., Liu, H. & Li, D. (2013). Iodine biofortification of vegetable plants - An innovative method for iodine supplementation. *Chinese Science Bulletin, 58,* 2066-2072.

Weng, H. X., Weng, J. K., Yan, A. L., Hong, C. L., Yong, W. B. & Qin, Y. C. (2008b). Increment of iodine content in vegetable plants by applying iodized fertilizer and the residual characteristics of iodine in soil. *Biological Trace Element Research, 123,* 218-228.

Whitehead, D. C. (1974). The sorption of iodide by soil components. *Journal of the Science of Food and Agriculture, 25,* 73-79.

Whitehead, D. C. (1981). The violatilisation, from soils and mixtures of soil components, of iodine added as potassium iodide. *Journal of Soil Science, 32,* 97-102.

Whitehead, D. D. C. (1975). Uptake by perennial ryegrass of iodide, elemental iodine and iodate added to soil as influenced by various amendments. *Journal of the Science of Food and Agriculture, 26,* 361-367.

WHO (2007). *Assessment of iodine deficiency disorders and monitoring their elimination: a guide for programme managers*, 3rd ed. Geneva: World Health Organisation.

Wong, G. T. F. (1982). The stability of molecular iodine in seawater. *Marine Chemistry, 11,* 91-95.

Wong, G. T. F. & Cheng, X. H. (2001). The formation of iodide in inshore waters from the photochemical decomposition of dissolved organic iodine. *Marine Chemistry, 74,* 53-64.

Xu, F. (1996). Catalysis of novel enzymatic iodide oxidation by fungal laccase. *Applied Biochemistry and Biotechnology, 59,* 221.

Yamada, H., Kiriyama, T., Onagawa, Y., Hisamori, I., Miyazaki, C. & Yonebayashi, K. (1999). Speciation of iodine in soils. *Soil Science and Plant Nutrition, 45,* 563-568.

Yamada, H., Kiriyama, T. & Yonebayashi, K. (1996). Determination of total iodine in soils by inductively coupled plasma mass spectrometry. *Soil Science and Plant Nutrition, 42,* 859-866.

Yasuo, M. & Shizuo, T. (1963). Evaporation of iodine from the ocean. *Journal of Geophysical Research, 68,* 3989-3993.

Yoch, D. C. (2002). Dimethylsulfoniopropionate: Its sources, role in the marine food web, and biological degradation to dimethylsulfide. *Applied and Environmental Microbiology, 68,* 5804-5815.

Yoshida, S., Muramatsu, Y. & Uchida, S. (1992). Studies on the sorption of I^- (iodide) and IO_3^- (iodate) onto Andosols. *Water, Air, and Soil Pollution, 63,* 321-329.

Yoshida, Y. M., Satoshi (1999). Effects of microorganisms on the fate of iodine in the soil environment. *Geomicrobiology Journal, 16,* 85-93.

Yuita, K. (1992). Dynamics of iodine, bromine, and chlorine in soil. *Soil Science and Plant Nutrition, 38,* 281-287.

Yuita, K. & Kihou, N. (2005). Behavior of iodine in a forest plot, an upland field, and a paddy field in the upland area of Tsukuba, Japan: Vertical distribution of iodine in soil horizons and layers to a depth of 50 m. *Soil Science and Plant Nutrition, 51,* 455-467.

Yuita, K., Tanaka, T., Abe, C. & Aso, S. (1991). Dynamics of iodine, bromine, and chlorine in soil: I. Effect of moisture, temperature, and pH on the dissolution of the triad from soil. *Soil Science and Plant Nutrition, 37,* 61-73.

Zhao, D., Lim, C.-P., Miyanaga, K. & Tanji, Y. (2013). Iodine from bacterial iodide oxidization by *Roseovarius* spp. inhibits the growth of other bacteria. *Applied Microbiology and Biotechnology, 97,* 2173-2182.

Zhu, Y. G., Huang, Y. Z., Hu, Y. & Liu, Y. X. (2003). Iodine uptake by spinach (*Spinacia oleracea* L.) plants grown in solution culture: effects of iodine species and solution concentrations. *Environment International, 29,* 33-37.

Zimmermann, M. B., Jooste, P. L. & Pandav, C. S. (2008). Iodine-deficiency disorders. *The Lancet, 372,* 1251-1262.

Zois, C., Stavrou, I., Kalogera, C., Svarna, E., Dimoliatis, I., Seferiadis, K. & Tsatsoulis, A. (2003). High prevalence of autoimmune thyroiditis in schoolchildren after elimination of iodine deficiency in northwestern Greece. *Thyroid, 13,* 485-489.

In: Advances in Environmental Research
Editor: Justin A. Daniels
ISBN: 978-1-53615-009-4

Chapter 2

USE AND APPLICATIONS OF IODIDE SALTS IN FUNGAL AND PLANT RESEARCH

Elena Fernández-Miranda[1,*,†] and Marcos Viejo[2,†]
[1]BioCarbon Engineering, Oxford, UK
[2]Department of Plant Sciences (IPV), Faculty of Biosciences (BIOVIT), Norwegian University of Plant Sciences (NMBU), Ås, Norway

ABSTRACT

The study of the microscopic structure of the biological material, which allows us to know how the unitary components are structurally and functionally related, has always been of paramount interest in the biomedical research. That knowledge is at the intersections between biochemistry, molecular biology and physiology on the one hand and pathological processes and their consequences on the other. Since the earliest years of the XIX century, several salts of iodide such as potassium iodide (Lugol) or more recently the Propidium iodide (PI) have been used in animal, plant, bacterial and fungal histology. Their application range goes from the detection of simple molecules to determine their content in a given cell or tissue (e.g., detection of

* Corresponding Author Email: cagigal_11@hotmail.com.
† Both authors contributed equal to this study.

polysaccharides in pollen of *Pinus pinea*), to the determination of physiological statuses (e.g., fungal spore viability of *Rhizopogon roseolus*). This work intends to provide a comprehensive overview of the uses of the iodide salts by especially focussing on its application in plants and microbiology, the mechanisms of action and the possible new uses in histology.

Keywords: histology, Lugol, Merzel, propidium iodide

INTRODUCTION

Iodine (I) is a non-metallic element belonging to the group of halogens (group 17 of the periodic table) and is the least abundant of the group (it is rarer than thulium, the most inaccessible among of all lanthanides).

The French chemistry Bernard Courtois discovered it in 1811 when he was trying to produce potassium nitrate (KNO_3), a substance high valued in at that time because it was used for obtaining gunpowder. In order to obtain KNO_3, he collected the waste seaweed (kelp) and then burned it, extracting the ashes with water, and producing a dark liquid that given the name of *soude de varech* or *salin de varech* (alkaline metals carbonates), which was purified by fractional crystallisation. The impurities were eliminated by heating the liquid with sulphuric acid. But one day, he added more acid than usual and when he warmed up the liquid; he noticed that a very conspicuous violet vapour was being emitted. The vapour condensed easily, leaving small, shiny black crystals.

$$2I^- + H_2SO_4 \rightarrow I_2 + SO_3^{2-} + H_2O$$

He did not have enough money to pursue the investigation, and he abandoned it, but he gave samples to Nicholas Clement and Charles Bernard Desormes, who, according to Partington (1964), passed the samples on to Louis Joseph Gay Lussac and Humphry Davy. Both men recognized that the discoverer of the substance was Courtois.

According to Martín–Sánchez et al. (2013), in 1813, Gay Lussac dedicated himself to studying those strange vapours formed out of a substance of metallic aspect, concluding it was a new element that he called 'iodine' [from the Greek iróSsç (ioeidés), or violet, for the colour of its vapour]. In the 1813 edition of *Annales de Chimie* (Davy, 1813), there are several articles that list all the properties of iodine and its derivatives. The author is undoubtedly Gay Lussac. But the most complete study data was published in 1814, when Gay Lussac made an exhaustive study of the action of iodine on all inorganic compounds, including the properties and behaviour of the products obtained, and he even created an appendix on the history of the discovery of iodine (Gay Lussac, 1813a, 1813b, 1814; Courtois, 1813).

In 1814, Stromeyer studied the action of iodine on inorganic bodies, and Colin and Gaultier studied its effects on organic bodies, demonstrating that iodine produces a blue coloration when it interacts with starch, according to Roscoe and Schoerlemmer (1884). In 1825, iodine began to be used as a plant histological stain when Raspail demonstrated the existence of starch granules in developing seeds (Lillie 1992). With slight modifications, the procedure is listed in all the textbooks on botanical micro techniques, quite often as a solution of iodine/potassium iodide (Johansen 1940), which is often called Lugol solution (Nemec 1962).

Iodine solutions are some of the most frequently-used chemical reagents in both mycology and lichenology. Iodine is able to produce colour reactions involving the formation of blueish (amyloid) and reddish (dextrinoid) complexes. Leonard (2006) reports that the earliest reference to using the bluing of fungi by iodine to identify fungi was in a report on the bluing of lichens in 1852, which was also noted by the Tulasne brothers (1861) in their great three-volume work on fungi in a chapter on the Seminis Fungini, 'The Fungus Seed'. They also noted the report of the bluing of an ascomycete *Amylocarpus encephaloides* Curr., by Currey in 1858 and they discussed bluing with sulphuric acid and iodine in young asci and ascospores. And Lillie (1992) described that the chitin is stained a reddish violet and/or violet also with iodine and $ZnCl_2$ when it is pre-treated with KOH.

For all these reasons, we can state that iodine and the different salts of iodide play a crucial role in the development of histochemistry and in the study of plants and fungi. In the current study, we will present the most relevant uses of the iodine and iodide in both groups.

LUGOL

Lugol (I_3K), a main representative of iodine salts, has proven to be an effective dye for detecting starch compounds in plant tissues. The Lugol reagent was used for the first time by Jean Guillaume Auguste Lugol (1829, 1830), a French doctor who used this solution at the Saint Louis Hospital in Paris to treat scrofula, which is an infectious process that affects the lymph nodes in the neck. Scrofula is caused by *Mycobacterium tuberculosis*.

Generally, lugol is associated with old-fashioned techniques and lab procedures, but nothing could be further from the truth. The localization of different metabolites and macromolecules is paramount in plant descriptive studies and in functional physiology studies. In a broad context, a high content and quality of starch in seeds has been pursued millennia because the survival of the human species depends upon it. The production of starch through crops allowed humans to settle and become the main source of energy. So the direct study of sugar metabolism by modern science and farmers not so far from the domestication and breeding make technology in the ancient times. Although sophistication has made the difference, the genuine interest is still the same.

Sugar is a primary metabolite of cell metabolism, and starch localization is as a good way to study sugar fluxes during normal plant development. For that reason reliable detection techniques have been demanded. The histochemical reaction of potassium iodide-iodine (KI - I_2) was the first one used.

Originally, several methods were developed for starch and cellulose identification by iodide salts (Jensen 1962). Importantly, iodine is used as a KI salt because of solubility. To understand the method, it is important to know in detail the composition and structure of starch: it is composed of variable amounts of 2 biopolymers: amylose [$(C_6H_{10}O_5)_n$] and amylopectin ($C_{30}H_{52}O_{26}$). While amylose is mainly a linear molecule, amylopectin is highly branched. Accumulation of both polymers depends on species origin, which in turn will determine the conformation of the starch granule (Buléon et al. 1998). Moreover, amylose tends to form a levogyre helicoidal structure with an internal cavity of 5 Å where iodine anions fit. Once formed, the starch-iodine complex absorbs light in the blue range conferring it this characteristic colour. Due to differences in starch composition, variations in the absorption range can be expected (Knutson 2000). Moreover, the starch-iodine complex may contain ions of different cationic species as claimed by Saenger (1984), although in recent years it has been proposed that iodine forms infinite polyiodide chains within the helix (Madhu et al. 2016).

In addition to the localization of starch in plant tissues, in 1999 Ernst and collaborators developed a histochemical procedure by fixing plant materials and Lugol staining after which they used microscopic images and image analysis. A finer starch content quantification using iodine-affinity capillary electrophoresis, which takes into account the ratio amylose-amylopectin, was developed by Herrero-Martínez et al. (2004).

In the 70's, iodine salts were adopted as a promising method for identifying polyglucosane accumulations (Kutík and Beneš 1977). In line started studies of glucose metabolism with an increased attention to the enzymatic processes behind it (Kahl 1973). Other studies focused on the importance of sugar fluxes as indicators of growth and *in vitro* organ differentiation (Thorpe and Murashige 1968; Kutík and Beneš 1979). Those early studies set the basis for coming years in which better approximations led to an exponential increase of the knowledge of sugar metabolism. Furthermore, iodine salts never disappeared from the laboratories due to their reliability, simplicity of use and low cost. In more recent work, iodine reagents have contributed to study alterations during the tuberization and starch deposition in potato (Kuipers et al. 1994) and the physicochemical characteristics of the starch (Flipse et al. 1996; Christensen and Madsen 1996).

Moreover, Lugol can be used to determine sugar remobilization in maple during natural leave senescence as a preparation for winter dormancy (Fulgosi et al. 2012). In *Arabidopsis* sp., a Lugol solution has been used to quantitatively determine the starch fluxes in the day-night cycle within chloroplasts (Crumpton-Taylor et al. 2012) and the amylopectin content in leaves (Zeeman et al. 2002).

Reproduction is a crucial step of plant ontogenesis. Gametogenesis and pollination are the first steps leading to fertilization and seed production. While gametogenesis generates genetic diversity, the successful pollination ensures the beginning of embryo and seed development, influencing fitness and giving rise to the next generation.

Assessment of pollen viability was always a main issue in agronomy research because the most important crops develop starch-containing pollen. Given its short-living nature, species greatly depend on accurate windows for pollen dispersion and rates of viability and longevity become of paramount importance to guarantee pollination efficiency (Wang et al. 2004). Thus, a technique to distinguish between viable and dead pollen is vital. Lugol staining is one such viability test. Pollen grains turn from light blue to black according to their starch concentration indicating viability. In many plant species and crops, knowledge on reproductive biology is scarce and fragmented, so in recent years there have been many efforts to establish reliable methods to better characterize important phenological traits. In fact, saccharides content, as an indicator of available reserves, can be quantified in mature pollen grains as shown by Speranza et al. (1997). In their work, starch content was quantified by image analysis after staining and microscopy.

Pollen studies are also valuable in ornamental breeding programs as for *Passiflora* spp. (Soares et al. 2013) when it comes to potential crosses between species. Soares et al. found inconsistencies in pollen viability assessment with Lugol *versus* production parameters (i.e., the number of fruits and seeds), so they suggest complementary *in vitro* germination testing to staining methods. Moreover, abiotic factors affecting pollen viability must be taken into account when controlled pollinations are required. According to this need, UV radiation, temperature and humidity seem to be the main parameters to control (Ge et al. 2011) as they influence pollen viability and might compromise the correlation with the iodide staining. Moreover, Lugol, along with other stains, has been used to assess the effect of pH on polysaccharide mobilization during pollen germination in *Pinus sylvestris*, and consequently pollen tube growth (Pardi et al. 1996).

Ecologically, the study of starch content in pollen grains has been linked with pollination type and plant evolution. In Baker and Baker (1979) there is a well-documented review on the topic. In there the authors completed a comprehensive evaluation of 990 species belonging to 124 families and quantified starch content (among other compounds) in pollen grains. Then, they could establish a correlation between morphometric parameters, nutritious reserves and its evolutionary connotations.

Lugol use is, however, not only restricted to plant histochemistry. Fungal tissue also has starch, even though according to Dodd and McCracken (1972) and McCracken (1974) it differs from plant starch. Usually, starch is considered being a storage carbohydrate. However, fungal starch does not appear to be a food reserve for two reasons: (1) as a cell wall component it is outside the spore or hyphal protoplast and therefore not readily available as an energy source, and (2) the starch molecules are short, unbranched chains. If starch was being stored as a nutrient reservoir, a branched product of higher molecular weight would likely be formed as in plants that store starch, or as in organisms that store glycogen.

In mycology, the term amyloid is currently applied to a blue iodine reaction of fungal microstructures. This colour reaction is due to an iodine complex with helicoid carbohydrate macromolecules. It is mainly observed when adding iodine reagents to ascus walls and ascus apical rings, but also to basidiospore walls and other fungal structures. Amyloidosis is a crucial characteristic in fungal taxonomy since around 1865. A special case of amyloid reaction is named hemiamyloid reaction. Baral (1987) defined hemiamyloid reaction as the red change in the apical tip of an ascus or the outer layer of the ascus wall produced by a Lugol solution. This peculiar colour reaction is so far only known in Ascomycetes, and of widespread occurrence particularly in lichens but also in many Helotiales and some Pyrenomycetes. The high taxonomic value of this red reaction can hardly be overestimated. But in mycology is not Lugol the iodide more crucial for taxonomy.

MELZER'S REAGENT

Melzer's reagent is an iodine solution employed for the study of Hymenomycetes, the vesicular-arbuscular mycorrhizal fungi spores, the asci of Ascomycetes, and other parts of fungi. It seems that even though iodine was used in the mid-1800s in lichen and asci evaluation, it was not until 1924 that the mycologist Václav Melzer documented iodine use for white spore identification. He described the usage of an iodine solution with chloral hydrate ($C_2H_3Cl_3O_2$) (50 g KI, 100 g $C_2H_3Cl_3O_2$, 1.5 g I_2, 100 ml dH_2O) to help show the spore ornamentation in the genus *Russula*. Chloral hydrate is a clearing, bleaching agent, thus improving the transparency of various dark-coloured microscopic materials and marking the difference among Lugol procedures.

A simple preparation can be made by adding fungal tissue or spores to a drop of Melzer's reagent on a microscope slide. One of the following reactions will be seen with the naked eye.

- Amyloid reaction: the material reacts blue to black. The amyloid reaction includes 2 subtypes of reactions:
  - Euamyloid reaction: test material turns blue without potassium hydroxide (KOH) pre-treatment.
  - Hemiamyloid reaction: test material turns red in Lugol solution, but shows no reaction in Merzel`s reagent, when pre-treated with KOH, it turns red in both reagents (Baral 1987)
- Dextrinoid reaction: Blackwell et al. (2001) suggested that the red reaction does not involve starch or amylose, but is instead a reaction with glycine betaine, an "osmolyte" (an organic osmotic solute) found in high concentrations in the Basidiomycetes under

study. Betaine is necessary to attract water to the rapidly enlarging and differentiating basidiomata. Iodine addition presumably results in a "glycine betaine-IKI complex." Baral (1987) noted that the dextrinoid reaction is "strongly enhanced" by chloral hydrate, as the material reacts brown to reddish-brown.

- Inamyloid: There is no colour change with Melzer's reagent, or the material reacts faintly yellow-brown.

One of the most common uses is the taxonomic classification of the spores of Glomalean fungus spores. The walls of these fungus spores have one or more layers of variable thickness, structure (Figure 1), appearance and staining reactions, and can be described using standardised terminology or diagrams ("murographs") (Walker 1983; Morton 1988). Iodine staining reactions will vary from pale pink (weak reaction) to dark red-brown (moderate reaction) to dark reddish-purple (intense dextrinoid reaction). Iodine binds to the hydrophobic regions of macromolecules (in spore or germinal walls) where the intensity of the reaction is partly related to the length of carbohydrate chains. In most instances to date, the intensity of the reaction is directly correlated to the plasticity of the structure in acidic mountants.

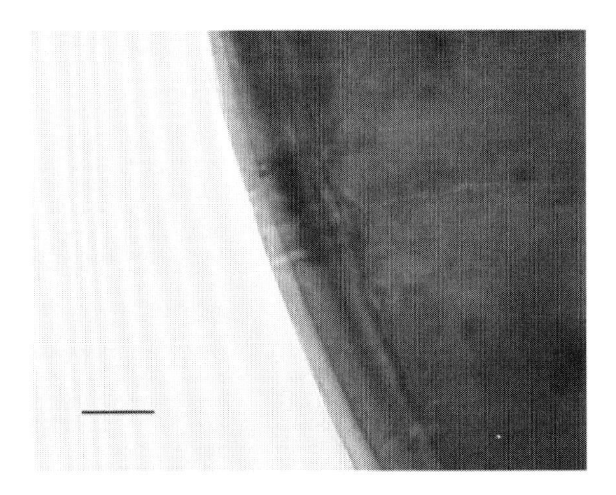

Figure 1. Wall of Glomeromycota spore (*Glomus* sp.) with Melzer's reagent showing the different layers in the wall.

PROPIDIUM IODIDE

Propidium iodide (PI) ($C_{27}H_{34}I_2N_4$) is a red fluorescent intercalating agent of nucleic acids with no base preference because it ionically binds to the bases´ phosphates. Therefore, it is used in cell nuclear staining. As PI binds stoichiometrically to nucleic acids, fluorescence emission is proportional to the nucleic acids content, which allows the quantification by several techniques. In aqueous solution, the dye has an excitation/emission maxima of 493/636 nm. Once the dye is bound, its fluorescence is enhanced 20- to 30-fold, the fluorescence excitation maximum is shifted ~30–40 nm to the red and the fluorescence emission maximum is shifted ~15 nm to the blue, resulting in an excitation maximum at 535 nm and fluorescence emission maximum at 617 nm.

Maybe the most striking feature of PI is its conditional ability of penetration of the cell membrane: only when the cell membrane integrity is compromised, becomes permeable to PI (Lehtinen et al. 2004). In combination with SYBR 14, it has been proven useful to distinguish between alive and dead zoospores (Stockwell et al. 2010). Conversely, 2-chloro-4-(2,3-dihydro-3-methyl-(benzo-1,3-thiazol-2-yl)-methylidene)-1-phenylquinolinium iodide (FUN-1) represents another alternative to assess viability and metabolic activity, as shown in yeast (Millard et al. 1997), urediniospores (Vittal et al. 2012), zoospores (Ivey and Miller 2014) and basidiospores (Fernandez-Miranda, 2017 Figure 2) and has been linked to apoptosis and programmed cell death studies (O´Brien et al. 1998).

Additionally, specific studies on plant cell death have taken advantage of PI´s features, both because of its fluorescence and its specificity on marking only cells with membrane damage. PI is, therefore, a remarkable programmed cell death staining. In a similar approach as Lugol, PI is well known as a viability marker in pollen (Regan and Moffatt 1990), even though more specific since it stains all dead pollen regardless of the starch content.

During plant development, there are several processes that require cellular death. For example, during root growth, it is necessary to selectively kill cells from the root cap in order to maintain the organ size

(Fendrych et al. 2014). Fendrych et al., used PI staining to mark cell death progression in the PCD (Programmed Cell Dead) sites within the root. In a similar context of normal development, PI staining of nuclei has been used to determine PCD involvement in the differentiation of tracheary elements of the xylem (Mittler and Lam 1995). In a novel approach, Jones et al. (2016) developed a protocol for confocal live cell imaging during fungus infection, which opens the possibility for more detailed time-series experiments regarding PCD in plants.

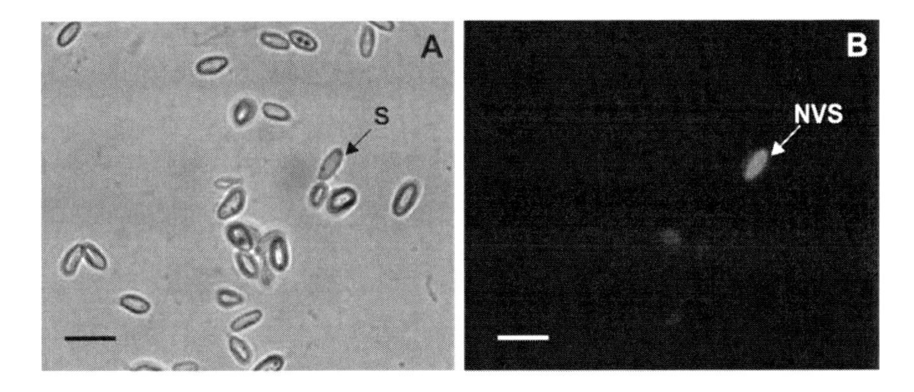

Figure 2. Spores of *Rhizopogon roseolus*. (A) observed under a light microscope; (B) the same spores under fluorescence microscopy stained with PI. In both cases, scale bar = 10 μm. S indicate Spore, NVS indicate non-viable spore.

However, the importance of PI does not end with its role as a DNA intercalant. It has also demonstrated to be a versatile compound since it also stains cell walls, so it is very appreciated as a structural counterstain in fluorescent microscopy studies Musielak et al. (2015). To which cell wall component PI binds is still not clear and the main hypothesis points to demethoxylated pectins (Rounds et al. 2011).

After studying 23 representative basidiomycetous yeast Zhang et al. (2018) found that some of these species (killed by 70% ethanol) could not be PI stained due to their complex cell wall structures. Based on these findings, they proposed that PI staining could serve as an indicator of cell wall complexity and also indicated that care must be taken when interpreting PI-negative cells as viable cells in the study of non-model microorganisms.

Since the 80's, PI has become a predominant compound in flow cytometry due to its mentioned characteristics; although it was already in use since the 70's when Bennett and Smith (1976) created the first nuclear DNA database in angiosperms by using flow cytometry and PI as staining. This was possible because of the direct proportionality between the fluorescence of the stained DNA and the amount of nDNA. In following years, the list, made for reference purposes, has been increased to over 1500 angiosperm species (Bennett et al. 2007); other authors also contributed to the increasing database with values on cultivated crops (Arumuganathan and Earle 1991). In those studies, the authors used flow cytometry to estimate genome size by measuring the fluorescence of stained nuclei (Doležel 1991). This data is of importance as the genome size is of great relevance in taxonomic studies (Murray 2005). Flow cytometry has also been used to identify ploidy level of variants in natural populations for population characterization and provenance selection for breeding (Johnson et al. 1998, Tuna et al. 2001).

In the general flow cytometry protocol by Galbraith et al. (1997), there are several precautions to be taken when using plant materials for nuclear DNA content analysis, such as the use of intact plant tissue or RNA removal to avoid interferences with the stained DNA measurements. As a standardized method, only leaves are used for general studies on a specific plant species. If different tissues from the same plant are analysed, variations on DNA size can occur (Bennet et al. 2007) due to inhibition by anthocyanins. Some cytosolic compounds may interfere with the measurements by inhibiting staining before flow cytometry; their nature and presence depend on environmental and genetic cues (Doležel and Bartoš 2005). Therefore, the selection of physiologically uniform tissues and inhibitor free is of utmost importance as shown by Price et al. (2000), who detected environmentally-induced inhibitors in leaves of *Helianthus annuus* affecting the intercalation and/or fluorescence of PI. Under the light of these results, the authors ask for a re-evaluation of previous studies.

In addition to the well-known relevance of PI in flow cytometry, it has been widely used in confocal imaging as a counterstain to identify nuclei in combination with GFP (Green Fluorescent Protein). In a comprehensive protocol (Truernit and Haseloff 2008) on several plant tissue types, the authors established the fluorescent emission to be collected for GFP and PI between 510 to 545 and 610 to 650 nm, respectively. Gathering data at these wavelengths has shed light on topics with profound implications for plant science such as parent-of-origin effects in embryo development (Bayer et al, 2009), hormone control in root meristem size (Ubeda-Tomás et al. 2009) or stem cell maintenance (Wildwater et al. 2005).

Acknowledgments

Sincere thanks to Abelardo Casares, PhD and Rosa García-Verdugo, PhD for their help and suggestions to this manuscript.

References

Arumuganathan, K. and Earle, E. D. 1991. "Nuclear DNA content of some important plant species nuclear DNA content material and methods." *Plant Molecular Biology Reporter.* 9, 208–218. doi: 10.1007/BF0 2672016.

Baker, H. G and Baker, I. 1979. "Starch in angiosperm pollen grains and its evolutionary significance." *Botanical Society of America.* 66, 591–600. doi: 10.1002/j.1537-2197.1979.tb06262.x.

Baral, H. O. 1987. "Lugol's solution/IKI versus Melzer's reagent." *Mycotaxon* Vol XXIX, pp. 399–450.

Bayer, M., Nawy, T., Giglione, C., Galli, M., Meinnel, T. and Lukowitz, W. 2009. "Paternal control of embryonic patterning in *Arabidopsis thaliana." Science.* 323, 1485–1488. doi: 10.1126/science.1167784.

Bennett, M. D. and Smith, J. B. 1976. "Nuclear DNA amounts in angiosperms." *Philosophical Transactions of the Royal Society B: Biological Sciences.* 274, 227–274. doi: 10.1098/rstb.1976.0044.

Bennett, M. D., Price, H. J. and Johnston, J. S. 2007. "Anthocyanin inhibits propidium iodide DNA fluorescence in *Euphorbia pulcherrima*: implications for genome size variation and flow cytometry." *Annals of Botany*. 101, 777–790. doi: 10.1093/aob/mcm303.

Blackwell, M., C. David, and S. A. Barker. 2001. "The presence of glycine betaine and the dextrinoid reaction in Basidiomata." *Harvard Papers in Botany* Vol. 6, No. 1, pp. 35–41.

Buléon, A., Colonna, P., Planchot, V. and Ball, S. 1998. "Starch granules: Structure and biosynthesis." *International Journal of Biological Macromolecules*. 23, 85–112. doi: 10.1016/S0141-8130(98)00040-3.

Christensen, D. H. and Madsen, M. H. 1996. "Changes in potato starch quality during growth." *Potato Research*. 39, 43–50. doi: 10.1007/BF02358205.

Courtois, M. B. 1813. "Découverte d'une substance nouvelle dans le Vareck." *Annales de Chimie*. 88, 304-310. ["Discovery of a new substance in the Vareck." *Annals of Chemistry*]

Crumpton-Taylor, M., Grandison, S., Png, K. M. Y., Bushby, A. J. and Smith, A. M. 2012. "Control of starch granule numbers in *Arabidopsis* chloroplasts." *Plant Physiology*. 158, 905–916. doi: 10.1104/pp.111.186957.

Currey, F. 1858. "On the existence of amorphous starch in a new *Tuberaceous fungus*." *Proceedings of the Royal Society of London*. 9, 119-23. doi: 10.1098/rspl.1857.0024.

Davy, H. 1813. "Sur la nouvelle substance découverte par M. Courtois dans le sel de Vareck." *Annales de Chimie*. 88, 322-329. ["On the new substance discovered by M. Courtois in the salt of Vareck." *Annals of Chemistry*.]

Doležel, J. 1991. "Flow cytometric analysis of nuclear DNA content in higher plants." *Phytochemical Analysis*. 2, 143–154. doi: 10.1002/pca.2800020402.

Doležel, J. and Bartoš, J. 2005. "Plant DNA flow cytometry and estimation of nuclear genome size." *Annals of Botany*. 95, 99–110. doi: 10.1093/aob/mci005.

Dodd, J. L. and McCracken, D. A. 1972. "Starch in fungi. Its molecular structure in three genera and a hypothesis concerning its physiological role." *Mycologia. 64*(6), 1341-1343. doi: 10.2307/3757973.

Ernst, M. K., Matitschka, G., Chatterton, N. J. and Harrison, P. A. 1999. "A quantitative histochemical procedure for measurement of starch in apple fruits." *The Histochemical Journal.* 31, 705–710. doi: 10.1023/A:1003992230135.

Fernández-Miranda, E., Majada, J. and Casares, A. 2017. "Efficacy of propidium iodide and FUN-1 stains for assessing viability in basidiospores of *Rhizopogon roseolus*." *Mycologia, 109*(2), 350-358. doi: 10.1080/00275514.2017.1323465.

Flipse, E., Suurs, L. C. J. M., Keetels, C. J. A. M., Kossmann, J., Jacobsen, E. and Visser, R. G. F. 1996. "Introduction of sense and antisense cDNA for branching enzyme in the amylose-free potato mutant leads to physico-chemical changes in the starch." *Planta.* 198, 340-347. doi: 10.1007/BF00620049.

Fendrych, M., Van Hautegem, T., Van Durme, M., Olvera-Carrillo, Y., Huysmans, M., Karimi, M., Lippens, S., Guérin, C. J., Krebs, M., Schumacher, K. and Novack, M. K. 2014. "Programmed cell death controlled by ANAC033/SOMBRERO determines root cap organ size in *Arabidopsis*." *Current Biology.* 24, 931–940. doi: 10.1016/j.cub.2014.03.025.

Fulgosi, H., Ježić, M., Lepeduš, H., Štefanić, P. P., Ćurković-Perica, M. and Cesar, V. 2012. "Degradation of chloroplast DNA during natural senescence of maple leaves." *Tree Physiology.* 32, 346–354. doi: 10.1093/treephys/tps014.

Galbraith, D. W., Lambert, G. M., Macas, J. and Dolezel, J. 1997. "Analysis of nuclear DNA content and ploidy in higher plants." *Current protocols in cytometry.* 2(1), 7-6. doi: 10.1002/0471142956.cy0706s02.

Gay Lussac, L. J. 1813. "Sur le nouvel acide formé avec la substance découverte par M. Courtois." *Annales de Chimie.* 88, 311- 318. ["On the new acid formed with the substance discovered by *M. Courtois*." *Annals of Chemistry.*]

Gay Lussac, L. J. 1813. "Sur la combinación de l´iode avec l'oxigène." *Annales de Chimie*. 88, 319-321. ["On the combination of iodine with oxygen." *Annals of Chemistry*.]

Gay Lussac, L. J. 1814. "Mémoire sur l´iode." *Annales de Chimie*. 91, 5-160. ["Memory on the iodine." *Annals of Chemistry*.]

Ge, Y., Fu, C., Bhandari, H., Bouton, J., Brummer, E. C. and Wang, Z. Y. 2011. "Pollen viability and longevity of switchgrass (*Panicum virgatum* L.)." *Crop Science*. 51, 2698–2705. doi: 10.2135/crop sci2011.01.0057.

Herrero-Martínez, J. M., Schoenmakers, P. J. and Kok, W. T. 2004. "Determination of the amylose-amylopectin ratio of starches by iodine-affinity capillary electrophoresis." *Journal of Chromatography A*. 1053, 227–234. doi: 10.1016/j.chroma.2004.06.048.

Ivey, M. L. L. and Miller, S. A. 2014. "Use of the vital stain FUN-1 indicates viability of *Phytophthora capsici* propagules and can be used to predict maximum zoospore production." *Mycologia*. 106:362–367. doi: 10.3852/106.2.362.

Jensen, W. A. 1962. *Botanical histochemistry: principles and practice*. San Francisco: W.H. Freeman & Co.

Johansen, D. A. 1940. *Plant Microtechnique*. McGraw-Hill Book Company, Inc.; London.

Johnson, P. G., Riordan, T. P., and Arumuganathan, K. 1998. "Ploidy level determinations in buffalograss clones and populations." *Crop Science*. 38, 478–482. doi:10.2135/cropsci1998.0011183X003800020034x.

Jones, K., Kim, D. W., Park, J. S. and Khang, C. H. 2016. "Live-cell fluorescence imaging to investigate the dynamics of plant cell death during infection by the rice blast fungus *Magnaporthe oryzae*." *BMC Plant Biology*. 16, 1–8. doi: 10.1186/s12870-016-0756-x.

Kahl, G. F. 1973. "Genetic and Metabolic Regulation in Differentiating Plant Storage Tissue Cells." *Botanical Review* 39, 274–279. doi: 10.1007/BF02860120.

Knutson, C. A. 2000. "Evaluation of variations in amylose – iodine absorbance spectra." *Carbohydrate polymers* 42, 65–72. doi: 10.1016/S0144-8617(99)00126-5.

Kuipers, A., Jacobsen, E. and Visser, R. 1994. "Formation and deposition of amylose in the potato tuber starch granule are affected by the reduction of granule-bound starch synthase gene expression." *Plant Cell.* 6, 43–52. doi: 10.1105/tpc.6.1.43.

Kutík, J. and Beneš, K. 1977. "Permanent slides after detection of starch grains with lugol's solution." *Biologia Plantarum.* 19, 309–312. doi: 10.1007/BF02923135.

Kutík, J. and Beneš, K. 1979. "Structural aspects of the regulation of starch accumulation in stem pith explants of kale." *Biologia Plantarum.* 21, 351–354. doi: 10.1007/BF02878232.

Lehtinen, J., Nuutila, J. and Lilius, E. M. 2004. "Green fluorescent protein–propidium iodide (GFP-PI) based assay for flow cytometric measurement of bacterial viability." *Cytometry Part A: The Journal of the International Society for Analytical Cytology*, *60*(2), 165-172. doi: 10.1002/cyto.a.20026.

Leonard, L. M. 2006. "Melzer's, Lugol's or Iodine for Identification of White-spored Agaricales?." *McIlvainea*, *16*(1), 43-51.

Lillie, R. D. 1992. *H. J. Conn's Biological Stains*. 2nd reprint of 9th ed. Sigma Chemical Company, St. Louis, MO, USA, pp. 692.

Lugol, J. G. A. 1829. *Mémoire sur l'emploi de l'iode dans les maladies scrophuleuses*, lu à L'Académie Royale des Sciences dans la séance du 22 juin 1829, 1, 78. J. B. París: Ballière. [*Memoir on the use of iodine in scrophulous diseases*, read at the Royal Academy of Sciences in the session of June 22, 1829,]

Lugol, J. G. A. 1830. *Mémoire sur l'emploi de l'iode dans les maladies scrophuleuses, suivi d'un tableau pour servir à l'administration des bains ioduré selon les ages.* 1, 52. París: J. B. Ballière. [*Memoir on the use of iodine in scrofulous diseases, followed by a table for use in the administration of iodized baths according to age.*]

McCracken, D. A. 1974. "Starch in Fungi: III. Isolation and Properties of an Amylose-precipitating Factor from *Lentinellus Ursinus* Fruit Bodies." *Plant physiology*, *54*(3), 414-415. doi: 10.1104/pp.54.3.414.

Madhu, S., Evans H. A., Doan-Nguyen, V. V. T., Labram, J. G., Wu, G., Chabinyc, M. L., Seshadri, R. and Wudl, F. 2016. "Infinite polyiodide chains in the pyrroloperylene–iodine complex: insights into the starch–iodine and perylene–iodine complexes." *Angewandte Chemie International Edition.* 55, 8032–8035. doi: 10.1002/anie.201601585.

Martín-Sánchez, M., Martín-Sánchez, M. T., and Pinto, G. 2013. "Reactivo de Lugol: Historia de su descubrimiento y aplicaciones didácticas." *Educación química*, *24*(1), 31-36. ["Reactive of Lugol: History of its discovery and didactic applications." *Chemical education.*]

Melzer, M. V. 1924. "L'ornementation des spores de Russeles." *Bulletin de la Société Botanique de France.* 40: 78–81 doi: 10.1080/00378941.1930.10833730. ["Ornamentation of Russeles spores." *Bulletin of the Botanical Society of France.*]

Millard, P. J., Roth, B. L., Truong Thi, H-P., Yue, S. T. and Haugland, R. P. 1997. "Development on the FUN-1 family of fluorescent probes for vacuole labelling and viability testing yeast." *Applied and Environmental Microbiology.* 63:2897–2905.

Mittler, R. and Lam, E. 1995. "In situ detection of nDNA fragmentation during the differentiation of tracheary elements in higher plants." *Plant Physiology.* 108, 489–493. doi: 10.1104/pp.108.2.489.

Morton JB. 1988. "Taxonomy of mycorrhizal fungi: classification, nomenclature, and identification." *Mycotaxon.* 32: 267-324.

Murray, B. G. 2005. "When does intraspecific C-value variation become taxonomically significant?" *Annals of Botany.* 95, 119–125. doi: 10.1093/aob/mci007.

Musielak, T. J., Schenkel, L., Kolb, M., Henschen, A. and Bayer, M. 2015. "A simple and versatile cell wall staining protocol to study plant reproduction." *Plant Reproduction.* 28, 161–169. doi: 10.1007/s00497-015-0267-1.

Němec, B. 1962. *Botanicka mikrotechnika.* [*Botanical Microtechnique*] Nakladatelstvi Ceskoslovenske akademie ved. Praha. [Publishing House of the Czechoslovakian Academy of Sciences, Prague - In Czech].

O'Brien, I. E. W., Baguley, B. C., Murray, B. G., Morris, B. A. M. and Ferguson, L. B. 1998. "Early stages of the apoptotic pathway in plant cells are reversible." *The Plant Journal.* 13, 803–814. doi: 10.1046/j.1365-313X.1998.00087.x.

Pardi, M. L., Viegi, L., Renzoni, G. C., Franchi, G. G. and Pacini, E. 1996. "Effects of acidity on the insoluble polysaccharide content of germinating pollen of *Pinus pinea* L. and *Pinus pinaster* Aiton." *Grana* 35, 240–247. doi: 10.1080/00173139609430010.

Partington, J. R. 1964. "*A history of Chemistry.*" vol. IV. London: MacMillan.

Price, H. J., Hodnett, G., Johnston and J. S. 2000. "Sunflower (*Helianthus annuus*) leaves contain compounds that reduce nuclear propidium iodide fluorescence." *Annals of Botany.* 86, 929–934. doi: 10.1006/anbo.2000.1255.

Regan, S. M. and Moffatt, B. A. 1990. "Cytochemical analysis of pollen development in wild-type Arabidopsis and a male-sterile mutant." *The Plant Cell.* 2(9), 877-889. doi: 10.1105/tpc.2.9.877.

Roscoe, H. E. and Schoerlemmer, C. A. 1884. *Treatise on Chemistry, Organic Chemistry.* Tomo 3, vol. 2. London: Macmillan.

Rounds, C. M., Lubeck, E., Hepler, P. K. and Winship, L. J. 2011. "Propidium iodide competes with ca 2+ to label pectin in pollen tubes and Arabidopsis root hairs." *Plant Physiology.* 157, 175–187. doi: 10.1104/pp.111.182196.

Saenger, W. 1984. "The structure of the blue starch-iodine complex." *Naturwissenschaften.* 71, 31–36. doi: 10.1007/BF00365977.

Sena, G., Wang, X., Liu, H. Y., Hofhuis and H. and Birnbaum, K. D. 2009. "Organ regeneration does not require a functional stem cell niche in plants." *Nature* 457, 1150–1153. doi: 10.1038/nature07597.

Soares, T. L., Jesus, O. N. de, Dos Santos-Serejo, J. A. and De Oliveira, E. J. 2013. "*In vitro* pollen germination and pollen viability in passion fruit (*Passiflora* spp.)." *Revista Brasileira de Fruticultura.* 35, 1116–1126. doi: 10.1590/S0100-29452013000400023.

Speranza, A., Calzoni, G. L. and Pacini, E. 1997. "Occurrence of mono- or disaccharides and polysaccharide reserves in mature pollen grains." *Sexual Plant Reproduction.* 10, 110–115. doi: 10.1007/s00 4970050076.

Stockwell, M. P., Clulow, J. and Mahony, M. J. 2010. "Efficacy of SYBR 14/propidium iodide viability stain for the amphibian chytrid fungus *Batrachochytrium dendrobatidis." Diseases of Aquatic Organisms.* 88:177–181. doi: 10.3354/dao02165.

Thorpe, T. A. and Murashige, T. 1968. "Starch accumulation in shoot-forming tobacco callus cultures." *Science.* 160, 421–422. doi: 10.1126/science.160.3826.421.

Truernit, E. and Haseloff, J. 2008. "A simple way to identify non-viable cells within living plant tissue using confocal microscopy." *Plant Methods.* 4, 1–6. doi: 10.1186/1746-4811-4-15.

Tulasne, L. R. and Tulasne, C. 1861. "Selecta Fungorum Carpologia 1 [Select Fungorum Carpologia 1]." *Erysiphei.* Paris, France: Typographie Impenale.

Tuna, M., Vogel, K. P., Arumuganathan, K. and Gill, K. S. 2001. "DNA content and ploidy determination of bromegrass germplasm accessions by flow cytometry." *Crop Science.* 41, 1629–1634. doi: 10.2135/cropsci2001.4151629x.

Ubeda-Tomás, S., Federici, F., Casimiro, I., Beemster, G. T. S., Bhalerao, R., Swarup, R., Doerner, P., Haseloff, J. and Bennett, M. J. 2009. "Gibberellin signaling in the endodermis controls Arabidopsis root meristem size." *Current Biology.* 19, 1194–1199. doi: 10.1016/j.cub.2009.06.023.

Vittal, R., Haudenshield, J. S. and Hartman, G. L. 2012. "A multiplexed immunofluorescence method identifies *Phakopsora pachyrhizi* urediniospores and determines their viability." *Phytopathology.* 102:1143–1152. doi: 10.1094/PHYTO-02-12-0040-R.

Walker, C. 1986. "Taxonomic concepts in the Endogonaceae. II. A fifth morphological wall type in endogonaceous spores." *Mycotaxon.* 25(1), 95-99.

Wang, Z. Y., Ge, Y., Scott, M. and Spangenberg, G. 2004. "Viability and longevity of pollen from transgenic and nontransgenic tall fescue (*Festuca arundinacea*) (Poaceae) plants." *American Journal of Botany.* 91, 523–530. doi: 10.3732/ajb.91.4.523.

Wildwater, M., Campilho, A., Perez-Perez, J. M., Heidstra, R., Blilou, I., Korthout, H., Chatterjee, J., Mariconti, L., Gruissem, W. and Scheres B. 2005. "The Retinoblastoma-related gene regulates stem cell maintenance in Arabidopsis roots." *Cell.* 123, 1337–1349. doi: 10.1016/j.cell.2005.09.042.

Zeeman, S. C. 2002. "Starch Synthesis in Arabidopsis. Granule Synthesis, Composition, and Structure." *Plant Physiology.* 129, 516–529. doi: 10.1104/pp.003756.

Zhang, N., Fan, Y., Li, C., Wang, Q., Leksawasdi, N., Li, F. and Wang, S. A. 2018. "Cell permeability and nuclear DNA staining by propidium iodide in basidiomycetous yeasts." *Applied microbiology and biotechnology. 102*(9), 4183-4191. doi: 10.1007/s00253-018-8906-8.

Biographical Sketches

Elena Fernández-Miranda Cagigal

Affiliation: BioCarbon Enginnering

Education: PhD

Research and Professional Experience: Degree in Biotechnology and posterior PhD in Plant Physiology and Mycology at the University of Oviedo (Spain). During years I studied the way of restore degraded and polluted areas as research in the University and protect ecosystems as Environmental Supervisor working for the government of Asturias (Spain). I currently at BioCarbon Engineering exploring the best way performance ecological restorations using drones and have to marry the biology and engineering.

Professional Appointments:

- Plant Scientist at BioCarbon Engineering. Oxford, UK. 08/2016 - Present

- Postdoctoral researcher in the Department of Plant Physiology, Department of Organisms and Systems, University of Oviedo. Oviedo, Spain. 11/2014-07/2016.
- Predoctoral researcher in the Department of Plant Physiology, Department of Organisms and Systems, University of Oviedo. Oviedo, Spain. 10/2008-10/2014.
- Environmental supervisor in the Government of the Principality of Asturias (Spain) developing environmental impact assessments and plans to restore degraded areas. 01/2009-10/2011.
- Research Fellow in the Regional Service for Agri-food Research and Development (SERIDA). University of Oviedo, Spain. Grao, Spain. 02/2007 - 01/2008.
- International Volunteer at the Charles Darwin Foundation, in Marine Conservation Area, working with green turtles. Galapagos Island, Ecuador. 03-06/2006.
- Internship in the Plant Health Laboratory (Phytopathology) of the Government of the Principality of Asturias. Oviedo, Asturias. 10/2005 - 01/2006.
- Internship in the Animal Production Area in SERIDA. Villaviciosa, Spain. 07-09/2005.

Honors:

- Scholarship courses 2000-2005 of the Ministry of Education of Spain for the realization of the Degree in Biology.
- Volunteer during the summer of 2006 within the project whale watching Asturian coast of the Principality of Asturias and Obra Social "La Caixa," which resulted in the book "Cetacean Asturian coast. Areas of conservation interest."
- Scholarship of the Vice-Chancellor of Students and Cooperation of the University of Oviedo for congress assistance (2007).

- Scholarship of the Vice-Chancellor for Academic Activities, Teaching and European Convergence of the University of Oviedo for enrolment of the rcsearch period of the third cycle (2007-2008).
- Guest lecturer in IES Universidad Laboral (29-04-2015).
- Review: Revista Iberoamericana de Micología, Pakistan journal of scientific and industrial research and Mycology.
- Postdoctoral Fellowship 2015 CONICET Argentina Resolución Nº 4885 date 10/12/2016.

Publications from the Last 3 Years:

- Coto, M., Casares, A., Fernández-Miranda, E. (2015). *Using the mycorrhizal fungus Scleroderma citrinum Pers. in the recovery of areas degraded by heavy metals.* Master Thesis. DOI: 10.13140/ RG.2.1.4825.2240.
- Fernández-Miranda, E. (2017). Dark Septate endophytes (DSE) in Polluted Areas. In: *Endophytic fungi: diversity, characterization and biocontrol*, (Ed) Hughes E.. pp. 125. Publisher: Nova Science Publishers, Inc., ISBN: 978-1-53610-341-0.
- Fernández-Miranda, E., Casares, A. (2017). Influence of the culture media and the organic matter in the growth of *Paxillus ammoniavirecens* (Contu & Dessì). *Mycobiology.* 45(3):172-177. DOI: 10.5941/MYCO.2017.45.3.0.
- Fernández-Miranda, E., Casares, A. (2017). Structure, biogenesis and nutricional dependence of *Paxillus ammoniavirescens sclerotia* (Boletales, Paxillaceae). *Anales del Jardín Botánico de Madrid* 74(1):e054. DOI: 10.3989/ajbm.2454. [*Annals of the Botanical Garden of Madrid*]
- Fernández-Miranda, E., Majada, J., Casares, A. (2017). Efficacy of Propidium Iodide and FUN-1 stain for viability germination in basidiospores of *Rhizopogon roseolus*. *Mycologia*. 109(2):350-358. doi: 10.1080/00275514.2017.1323465.

Marcos Viejo

Affiliation: Norwegian University of Life Sciences. Faculty of Biosciences. Plant Science Department.

Education: PhD

Research and Professional Experience: My research has focused on plant epigenetics and physiology in forestry species, mainly European chestnut and Norway spruce. I have experience on *in vitro* culture and molecular biology techniques. I have collaborated in national and international projects and teaching as a PhD student.

Professional Appointments:

- 2008-2015: PhD student and researcher in several national and international projects. Spain
- 2015-present date: Postdoc researcher in the Norwegian University of Life Sciences. Norway

Publications from the Last 3 Years:

Carneros E, Yakovlev I, Viejo M, Olsen JE, Fossdal CG (2017). The epigenetic memory of temperature during embryogenesis modifies the expression of bud burst-related genes in Norway spruce epitypes. *Planta*, 1-14.

Pérez M, Viejo M, LaCuesta M, Toorop PE, Cañal MJ (2015). Epigenetic and hormonal profile during maturation of *Quercus suber* L. somatic embryos. *Journal of Plant Physiology*, 173:51-61.

In: Advances in Environmental Research
Editor: Justin A. Daniels
ISBN: 978-1-53615-009-4

Chapter 3

FLORISTIC, PHYTOSOCIOLOGY AND ECONOMIC POTENTIAL OF PLANT SPECIES FROM A SECTION OF VEGETATION IN THE NORTH OF PIAUÍ STATE, NORTHEASTERN BRAZIL

***Lucas Santos Araújo*[1], *Graziela de Araújo Lima*[2] *and Jesus Rodrigues Lemos*[3]**
[1]Biological Sciences from
the Federal University of Piauí/*Campus* Ministro Reis Velloso-CMRV,
Parnaíba, Piauí, Brasil
[2]Science and Biology teacher of Elementary Education
[3]Biological Sciences,
Federal University of Piauí/CMRV, Parnaíba, Piauí, Brasil

Abstract

This study aimed to contribute to a better knowledge of Piauí´s vegetation through a floristic survey, phytosociological study and knowledge about the economic potential and geographic distribution of species from an area belonging to the municipality of Brasileira, north of Piauí, Brazil. Fifty five families, 126 genera and 141 species represented the flora. The families with the highest species richness were Fabaceae, Lamiaceae, Malvaceae, Apocynaceae and Bignoniaceae. In the study area, 35 species, according to "Flora of Brazil 2020," were registered as endemic to Brazil. The species that had the highest importance values (VI) in the studied plant community were *Ephedranthus pisocarpus*, *Copaifera langsdorffii*, *Myrcia guianensis*, *Terminalia fagifolia*, *Parkia platycephala*, *Astrocaryum campestre* and *Anacardium occidentale*. The Shannon Index (H ') was 2.85 nats ind.-1 and the Equability Index (J') was 0.696. The maximum, mean and minimum heights of the individuals were 15m, 7.6m and 3.4 m and the maximum, average and minimum diameters were 47.7 cm, 9.2 cm and 5 cm, respectively. Fifty seven species distributed in 32 families present economic potential, being the majority of medicinal and meliferous use, being *Byrsonima correifolia*, *Caryocar coriaceum*, *Handroanthus impetiginosus*, *Magonia pubescens* and *Parkia platycephala* the ones with greater number of use in the literature. *Combretum leprosum*, *Cereus jamacaru* and *Bauhinia ungulata* were widely distributed in all plant formations compared to the local flora.

Keywords: flora, plant community, use of species, geographic distribution

Introduction

Brazil's rainforest has one of the largest territorial extensions in the world, with continental dimensions of approximately 8,500,000 km², with the richest biodiversity on the planet, having approximately 13% of the world's total biota, ranking first in number of plant species. It is estimated that the Brazilian flora is represented by 42,903 species (Ribeiro and Walter, 1998; Barbosa and Peixoto, 2003; Fernandes, 2003; Brandon et al., 2005; Lewinsohn and Prado, 2005; Forzza et al., 2012); sheltering seven biomes: Amazon, Cerrado, Pantanal, Atlantic Forest, Caatinga, "Campos Sulinos" and the coastal biome.

The Cerrado is the second largest Brazilian biome, occupying 21% of the national territory and comprises the set of ecosystems such as savannas, forests, grasslands, wetlands and gallery forests (Eiten, 1977; Ribeiro et al., 1981), and can be found in the states of Goiás, Tocantins, in a portion of the states of Bahia, Maranhão, Mato Grosso, Mato Grosso do Sul, Minas Gerais, Piauí and Rondônia, as well as certain areas in the northern states of Roraima, Pará, Amazonas, and delimited areas in the south of São Paulo and Paraná (Mesquita and Castro, 2007).

This biome concentrates one-third of the national biodiversity and 5% of the world's flora and fauna (Faleiro et al., 2008). Thus, it is a holder of the mega diversity of the planet, with about 40% of its endemic species (Lima et al., 2010), with a great diversity of habitats, which determine a remarkable alternation of species among different phytophysiognomies (Klink and Machado, 2005).

Besides the environmental aspects, the Cerrado has great social importance, many populations survive from its natural resources, including indigenous ethnic groups, geraizeiros, riparian, babassu palm, tidewater settlers and quilombola communities that, together, are part of the Brazilian historical and cultural patrimony and own a traditional knowledge of its biodiversity. Two Thousand and twenty plant species, and more, have medicinal uses and more than 416 can be used to recover degraded soils, such as wind barriers, erosion protection, or to create a habitat for natural pest predators (MMA, 2018).

The Cerrado, although having ecological, historical and cultural importance in Brazil and a very high biodiversity, Castro (1998) comments that this biome is suffering great anthropic interference. After the Atlantic Forest, the Cerrado is the Brazilian biome that suffered the most changes with human occupation. The two main threats to biodiversity in the Cerrado are related to two economic activities: intensive grain monoculture and extensive low-tech livestock farming. The use of techniques of intensive use of the soil has caused, for years, the exhaustion of the local resources. The indiscriminate use of agrochemicals and fertilizers has also contaminated soil and water (WWF-BRASIL, 2018).

In addition, WWF-Brazil (2018) argues that habitat destruction and fragmentation are currently the greatest threat to the integrity of this biome: 60% of the total area is devoted to livestock and 6% to grains, mainly soybeans. In fact, humankind due to agricultural expansion, urban expansion and road construction has already modified about 80% of the Cerrado - approximately 40% partially retain its initial characteristics and another 40% have totally lost them. Only 19.15% correspond to areas in which the original vegetation is still in good condition. In this scenario, the accelerated destruction of Cerrado native vegetation formations, together with the high diversity of endemic species, has led to the inclusion of the Cerrado among the global biodiversity hotspots (Myers et al., 2000; Silva and Bates, 2002).

The Brazilian Northeast is the one that presents the greatest diversity of natural pictures, which are revealed in its climate, vegetation and soil. According to Andrade (1977), the climate ranges from the superhumid to the semiarid and a mosaic of soil types, ranging from the mature zonals, nutrient poor by intense leaching to those of reduced alteration and high stock of minerals. In terms of vegetation, humid forests, forest river, Cerrado, hypoxerophilic and hyperxerophilic Caatingas and disjunctions between two plant formations (Foury, 1972).

Within this context, in the Northeast, the Piauí and Maranhão states have the largest area of Cerrado vegetation, occupying the southwest and central-north (Piauí) and south-central and northeast (Maranhão) portions, mainly in the portion corresponding to "Parnaibana Basin", designated by Castro (1994), of "distal marginalized Cerrados," since they are distributed in the margins of the phytogeographical space occupied by the Cerrados of Brazil, composing a northern extension of the central Cerrado.

The Piauí state has a total area of 25,093,400 ha, of which approximately 33.3% is covered by Cerrado sensu lato and 14% by transition areas between Cerrado and other types of vegetation: Caatinga, Carrasco, dry deciduous forest, semideciduous forest, "Mata dos Cocais forest, etc. demonstrating high floristic richness due to its unique geographic position, inserting itself in an area of ecological tension (ecotone), under a complex mosaic of vegetation types. For the Cerrados do Piauí the bibliographic base is still small (CEPRO, 1996; Farias and Castro, 2004; IBGE, 2004).

Due to the great ecological and historical-cultural importance of the Cerrado in Piauí, scientific studies of biodiversity are necessary, since, also, anthropic actions and the acceleration of the fragmentation and destruction of habitats have been occurring increasingly and generated worries. In this perspective, the present research intends to contribute with a greater collection of information, from the accomplishment of a floristic survey, phytosociological study and knowledge of the economic potential of the vegetal species of a Cerrado area located in Brasileira city, central-north of Piauí. These data could stimulate further studies, offering a wider range of information on the flora of this State, besides subsidizing actions of natural resource management programs and being adopted in public conservation policies of this region.

Literature Review

Brazil is considered one of the most diverse biologically countries because it houses about 10% of the living forms on the planet (Myers et al., 2000). The most abundant country in terms species' number, Brazil shows the largest portion of tropical forests in the world (Barbosa et al., 1996; Ribeiro and Walker, 1998). The Brazilian floristic diversity of Angiosperms is underestimated by 85,400 species, as evidenced by the Brazil Flora Group (BFG, 2015).

It is considered that a national, regional and local scale the surveys of floristic and phytosociological surveys have an undeniably illusory number due to the scarcity of this work and professionals able to carry them out. As a reference we have the flora of Northeast Brazil that is calculated in about 25,805 species of Angiosperms (BFG, 2015).

Despite Brazil having a poorly referenced vegetation, Giulietti et al. (2009) aiming to know the most important regions for the preservation of species and to map the endemic species of Brazilian plants, mention that for the country there are 2,256 rare species and 752 key areas for biodiversity.

The initial studies in the Northeast region were indicated from the precursor of Dárdano de Andrade Lima, still in the 1950s (Santos-Filho, 2009). Despite the fact that an important floristic diversity is observed in the plant communities of the Northeast region (Zickel et al., 2004), there is a great lack of published works, with only isolated floristic surveys (Andrade-Lima, 1951; Cabral-Freire and Monteiro, 1993).

In the Northeast, several studies have been carried out so far, with different purposes. Between the pioneer and classic stand out the projects of Tavares et al. (1969a, b, 1970, 1974a, b, 1975), which, through SUDENE (Superintendency for the Development of the Northeast), carried out forest inventories in the Caatinga to describe, characterize and evaluate the logging potential of xerophytic forests (Sampaio, 1996).

Because of its geographical location, the state of Piauí has a vegetation with influence of different domains: Amazonian, Central Plateau and semi-arid Northeastern. This is due to its characteristic climate and its geological formations (Farias and Castro, 2004).

However, from this great variety and in the past have been visited by influential naturalists such as Martius (1819), Gardner (1839), Schwacke (1878), Taubert (1895), Luetzelburg (1912/1914) and Assis Iglesias (1914-1919), the vegetation in Piauí does not have a detailed study, glimpsing an overall classification (Baptista, 1981).

Santos-Filho (2009) comments that in the 70's, scientists made extensive excursions with a scientific-exploratory method, starting the first attempts to classify the Piauí vegetation. Fernandes (1982) saw the geographic arrangement, organized the vegetation of Piauí into seven groups: forests, "Mata dos Cocais" forest ("carnaubais," "babaçuais" and "buritizais"), Cerrado (tropical savanna vegetation), Caatinga (thorn deciduous vegetation), carrasco (non-thorn deciduous lands), grassy field vegetation and coastal vegetation.

Santos-Filho (2009) also comments that few bibliographic works and technical reports without scientific character summarized the data of the flora of Piauí and that appeared in the 80's, the first academic researches (master dissertations and doctorate thesis) releasing information about the formation and organization of the vegetation of Piauí.

The following will be mentioned floristic and phytosociological surveys carried out, so far, in the different plant formations of the Piauí state.

Floristic Studies in Piauí´s State

Barroso & Guimarães (1980), in the Sete Cidades National Park, carried out a broad and pioneering floristic survey and listed 229 species, including herbaceous, sub-shrub, lianas, shrubs and trees

Oliveira et al. (1997) carried out a survey of the flora and phytosociological study of an area of vegetation of Carrasco-Caatinga transition in the municipality of Padre Marcos. In this study, Bignoniaceae was the most well represented family, followed by Fabaceae.

Farias (2003) carried out a floristic survey in Campo Maior verified that the vegetation of the study area, when compared to other surveys conducted in several vegetation types of the Northeast of Brazil, regarding the flora, was more related to the cerrado, followed by the vegetation of carrasco and caatinga.

Lemos (2004), studying the floristic composition of Serra da Capivara National Park docketed a total of 210 species distributed in 149 genera and 62 families. The families with the greatest wealth were Fabaceae, Bignoniaceae, Euphorbiaceae and Myrtaceae. The author also verified that the study area has similarity to other areas of the semi-arid Northeast caatinga, whether they are installed in sedimentary lands or on the crystalline basement.

Oliveira et al. (2007) carried out the floristic survey of the species occurring in the Paquetá Environmental Park, Batalha, verified that this municipality is characterized by typical formations of transition area between the cerrado and caatinga, and, in some places, riparian forests. They identified 109 species distributed in 37 families and 92 genera. The families that stood out most in the study area were Fabaceae, Asteraceae, Bignoniaceae, Rubiaceae and Combretaceae.

Matos (2009) in a study carried out in the Sete Cidades National Park found that the gallery forests presented high richness and floristic diversity, possibly due to the fact that the area is an environmental protection site. It was recorded 75 tree species distributed in 56 genera and 30 botanical families. The number of species sampled in this survey added 29 new tree species to the already existing listing for the Park.

Santos-Filho (2009), studying the floristic and structural composition of the Restinga vegetation in Piauí state, observed that the Angiosperm flora, until the moment of its research, was represented by 213 species belonging to 53 botanical families. The most representative families were Fabaceae, Euphorbiaceae, Bignoniaceae, Malvaceae, Myrtaceae, Rubiaceae, Amaranthaceae, Apocynaceae, Convolvulaceae and Cyperaceae, which together corresponded to 58.29% of the total recorded species.

Costa (2010), in a study carried out in the deltaic plain of the Parnaíba River (Piauí/Maranhão), aimed to contribute to the study of the Phytogeography of the Delta of the Parnaíba River, inserted in the states of Piauí and Maranhão. The results showed that the coastal vegetation is an extensive area of species with different stratifications and a significant morphological degree and physiological heterogeneity.

Silva et al. (2014) performed a study of the apicultural flora in two municipalities with great potential in honey production in Caatinga areas with the objective of identifying the botanical families and the species visited by bees. The occurrence of 35 melliferous species was recorded for the studied areas, distributed in 19 botanical families, the most representative being the families Malvaceae (six spp.) and Euphorbiaceae and Leguminosae (five spp. each).

Amaral and Lemos (2015) carried out a floristic survey in the municipality of Luiz Correia, in the coastal region of the Piauí state. The authors, taking into account the floristic lists of different plant formations present in the northeastern semi-arid region, found that the flora present in the studied area has elements present in the caatinga, cerrado and restinga, suggesting that the plant community located in the studied section of the coastal zone of Piauí has a transitional nature.

Sousa, Araújo and Lemos (2015) carried out a study with residents in the municipality of Buriti dos Lopes, Piauí´s state. It aimed to raise the plant species known and used by the native population for different purposes. Sixty taxa were identified, distributed among 52 genera and 27 families. The families with the highest number of species were Fabaceae (16 species), Combretaceae (four species), Anacardiaceae, Apocynaceae, Myrtaceae and Rubiaceae (three species each).

Rocha et al. (2017) carried out a floristic survey in an area of the Caatinga in the Gameleira do Rodrigues, located in the municipality of Picos, central state. They recorded the tree composition and verified the similarity of the flora with other studies carried out in the Brazilian Northeast. They observed that 9.7% of the species were exclusive to the study area, and the most expressive families in the number of species were Fabaceae (13 spp.), Anacardiaceae (four spp.) and Sapindaceae (three spp.), corresponding to 54, 05% of the total collected species. The floristic similarity revealed greater affinity to the vegetation found in Iguatu, Ceará´s state and São José do Piauí, Piauí.

Santos et al. (2017) with the objective of creating a dichotomous key of the type, based on vegetative characters, prepared a list for the arboreal and shrub species of a stretch of urban vegetation in the municipality of Parnaíba, Piauí, in which 33 species belonging to 21 botanic families. The most representative families in number of species were Fabaceae (10 spp.) and Rubiaceae (four spp.). The predominant habit of life was shrub.

Silva, Silva and Sousa (2017) carried out a floristic inventory of the vegetation occurring in Nazareth Eco Resort, municipality of José de Freitas, Piauí, comparing the study area with cerrado areas in Piauí. It was observed a heterogeneous flora, due to the influence of other plant formations, considered "Marginal distal Cerrados." In the floristic inventory were registered 46 families, 105 genera and 139 species. The families with the highest number of species were Fabaceae (31), Malvaceae (11), Rubiaceae (8), Asteraceae (7), Cyperaceae and Malpighiaceae (6) and Combretaceae (5). These families contributed approximately 52% of the total sampled species.

Phytosociological Studies in Piauí State

Studies of phytosociological parameters in Piauí are restricted mainly to vegetation areas of Cerrado and Caatinga, with predominance for the first one (Lima, Teodoro and Lemos, 2018). Castro (1994) did the early studies, who made a phytosociological comparison between two areas of "Cerradões Marginais," one in Oeiras, Piauí and another in Santa Rita do Passo Quatro, São Paulo. The phytosociological comparison between the two areas revealed 10 species, 31 genera and 22 families common to both taxocenoses.

Rodrigues (1998) showed vegetation in the municipality of Gilbués through transects with plots in a systematic way, in an area of desertification. A total of 126 species were recorded: 41 in the shrub-tree stratum and 55 in the herbaceous-sub-shrub stratum.

Lemos and Rodal (2002) carried out the phytosociological survey of the woody component of a section of the Caatinga vegetation in a sedimentary plateau area of Serra da Capivara National Park. They concluded that thc studied area presented greater floristic similarity with other Caatinga areas present in sedimentary basements, especially of the Mid North basin.

Mendes (2003) made the characterization of the floristic composition, architecture and structure of a fragment of arboreal Caatinga of the vegetation in Morro do Baixio, São José do Piauí. The vegetation of the studied area was framed as an arboreal Caatinga, with high frequency of typical species of sedimentary environments. The author stated that abiotic conditions such as temperature, precipitation and altitude condition a larger vegetation size in the studied area when compared to the deciduous vegetation of crystalline terrains.

Farias and Castro (2004) studied the floristic composition and the phytosociological structure vegetation of "Campo Maior Complex" and stated that the area is characterized as an environment subject to frequent floods, conferring the character of transition, physiognomic changes in its composition and in the arrangement of the species, passing from Campo to several physiognomic types of Cerrado, Caatinga, Carrasco and Semi-deciduous Forest.

Oliveira (2004) contributed to the knowledge of the flora and structure of Brazil´s Marginal Cerrados, in the Sete Cidades National Park, municipalities of Brasileira and Piracuruca (Parnaíba River basin). For the phytosociological survey, 73 plots of (100 m^2 each) were installed, registering 2,516 individuals, belonging to 139 species and 36 families.

Albino (2005) carried out a floristic and structural survey of a rupestrian cerrado ("cerrado" a bedrock vegetation) of low altitude in the municipality of Castelo of Piauí. In 10 sample units of 20m x 50m, 829 individuals belonging to 22 species and 15 families were sampled.

Costa (2005) performed the first qualitative-quantitative survey of the flora and quality of the melissofauna in the 'Percurso da Raposa' locality, in Castelo do Piauí. The author used the quadrant method, allocating 100 sample points distributed at 10m intervals. He considered the analysis satisfactory, in order to estimate the population size of the species that stood out as melitophilous pasture and some parameters revealed a phytophysiognomies with low plant diversity in the rock field.

Mesquita and Castro (2007) characterized floristic and vegetation structure of "Cerrado Marginal" ("Cerrado Baixo") in the Sete Cidades National Park. They concluded that the vegetation presents enormous physiognomic and floristic variety with areas of Cerrado *sensu stricto*, "Campo Cerrado" and "Campo Limpo."

Haidar (2008) investigated the diversity and structure of the tree community in three seasonal forests of the Cerrado biome, two in the Central Plateau (Goiás and Federal District) and one in the Parnaibano Sector (Piauí). Twelve species were found common to the seasonal forests of the Cerrado biome, including the two forests of the Central Plateau and one of the Sete Cidades National Park. They showed high diversity among the three seasonal forests and greater floristic similarity between the two areas of the Central Plateau.

Castro et al. (2009) investigated the transitional vegetation of the Serra Vermelha plateau area, in the municipalities of Redenção do Gurguéia, Bom Jesus, Morro Cabeça no Tempo and Curimatá, with applications of the Minimum Phytosociological Assessment Protocol (PAFM). The Transitional Semideciduous Seasonal Forest of the Serra Vermelha plateau area presented greater similarities with the "Sedimentary Caatinga" and the "Carrasco," moving away from the framework as the Atlantic Forest of the Northeast.

Sousa et al. (2009) developed a study in the "Serra de Campo Maior" (Serra de Santo Antonio, Serra de Bugarim and Serra de Passa-Tempo), Campo Maior, Piauí, with the objective of characterizing the vegetation of the area through floristic and phytosociological surveys, which was used the 20x50m plots method, with 30 plots allocated and a total of 2,646 individuals distributed in 31 families, 51 genera, 63 species and a totally unknown species. The most representative families were: Vochysiaceae (102.45%), Malpighiaceae (47.13%), Erythroxylaceae (33.01%), Mimosaceae (24.95%), Myrtaceae (22.89%) and Euphorbiaceae (13.86%).

Matos and Felfili (2010) listed the floristic composition, phytosociology, diversity, natural regeneration and relationship with environmental variables of the vegetation of the Matas de Galerias National Park in Sete Cidades National Park, as well as the floristic similarity between these forests and several other localities of the Cerrado biome. The vicinities analyzes showed that there is a great environmental heterogeneity in the area and a diverse flora composed by many species unique to the region.

Lindoso et al. (2010) sampled, in 10 plots of 20m x 50m, the floristic, structure and diversity of the Cerrado sensu stricto on soil quartz-sand neosol in spots in the Sete Cidades National Park. There were 45 species distributed in 21 families. The area presented density of 1,017 individuals.ha^{-1} and basal area of 10.71 m^{2}.ha^{-1}. The authors concluded that the density, basal area, richness and alpha diversity are within the limits found in the Cerrado sensu stricto in the general biome, indicating similarity of the "Cerrado marginal."

Moura et al. (2010) provided data on the floristic composition, diversity and structure of the woody component of the Cerrado sensu stricto on rock outcrops in the Sete Cidades National Park and compared the results obtained with other studies in the region and with other areas of Cerrado on rocky soils in Brazil Central. They located 10 plots of 20m x 50m at random. The study recorded 47 species, belonging to 38 genera and

20 families. The most representative families regarding the number of species were Fabaceae, Bignoniaceae and Vochysiaceae.

Haidar et al. (2010) evaluated the floristic, phytosociology and diversity of the natural stains of Semideciduous Seasonal Forest of the Sete Cidades National Park. Twenty-five plots of 20m x 20m were distributed among the three patches that occurred in the area. The floristic and structural similarity indices indicate differences between and within the Seasonal Forest Stains sampled in the Park, reflecting the heterogeneity of the soils and, consequently, the high beta diversity within the same physiognomy along the geographical space of the Park.

Lima et al. (2010) in his survey of floristic composition and phytosociological a Savannas community an area adjacent to the Sete Cidades National Park, in 17 plots of 20m x 30m at the base of "Morro do Cascudo," sampled 2,146 individuals grouped into 27 families and 71 species. The authors concluded that the denser the area, the greater the richness in relation to the number of plant species and that the biological invasion and degree of ecotonicity are not conspicuous, and cannot consider the area as fully ecotonal.

Pessoa and Santos-Filho (2011) differentiated the flora and the structure of the herbaceous flora in five areas (Capitão Gervásio de Oliveira, São João do Piauí, Brejo do Piauí, Elizeu Martins and Ribeiro Gonçalves) distributed in the semi-arid region of the state of Piauí, where transects of 5m each were installed, perpendicular to the edge (road side). Among the five different phytophysiognomies studied, Capitão Gervásio was the one with the highest species richness and the others presented very similar riches.

Amaral et al. (2012) studied floristic and phytosociology in a Cerrado-Caatinga transition area in the municipality of Batalha. They recorded 1,801 individuals, 34 species in 34 genera, 20 families. Although the area presents a relatively low number of species, the diversity found in the area was very low.

Mendes et al. (2012) evaluated the herbaceous-sub-shrub stratum in "Campo limpo" communities in the Sete Cidades National Park. It was sampled 71 taxa related to 46 genera and 25 families. Based on the results found in the mentioned study, they suggested that there is an association of vegetation with the edaphic characteristics in the Park. The soils of the six areas, although with low fertility and acid pH, presented differences in the contents of some elements and in the texture that dictated changes in the composition and structure of the species.

Alves et al. (2013) described data on the vegetation structure in an area of Caatinga in the municipality of Bom Jesus. The authors concluded that the diversity found in the area was considered median when compared to other areas of the caatinga.

Lima et al. (2013) recorded the physiognomy and vegetation structure of a "Cerradão" fragment from the Experimental Farm of the Agricultural College in Floriano. They found that the study area is preserved and presents a characteristic flora of "Cerradão," however with low species richness, density and total basal area.

Castro et al. (2014), characterized the tree community structure of a remnant of Seasonal Transitional Forest, with a decidual and a less dry, semidecidual part, in the municipality of Manoel Emídio and Alvorada do Gurguéia. The research referred to the diagnosis of the vegetation cover of Fazenda Novo Mundo and surroundings, with the purpose of providing a regional flora and phytosociology prognosis, associated with the environmental conditions prevailing in the mentioned municipalities. The authors installed 17 plots of 20m x 30m, sampling 4,276 woody individuals. Of the 78 specimens, 38 were identified at the species level, 25 at the gender level, and 14 at the family level. Fabaceae showed greater richness (12), followed by Bignoniaceae and Myrtaceae (10 species each).

Silva et al. (2015) carried out a floristic, phytosociological and ecological study of riparian forest located in the Gurguéia river basin in the municipality of Bom Jesus, south of Piauí, in a cerrado-caatinga transition area. In the vegetation analysis, a fixed plot method was used systematically; five tracks of 1200 m^2 (20m x 60m) were subdivided into three subunits of 20 x 20m, totaling 15 sample units.

The most recent floristic and phytosociological studies carried out and published in Piauí were by Vasconcelos et al. (2017), in which they knew the floristic-phytosociological diversity and the timber production of a caatinga area destined to the sustainable forest management; that of Lima, Teodoro & Lemos (2018) in which the authors structurally characterized a sub-deciduous vegetation stretch of a caatinga-cerrado transition area in the Brazilian Agricultural Research Company-EMBRAPA MEIO NORTE/Parnaíba Research Execution Unit; of the Silva & Lemos (2018), where an inventory of the vegetation cover was carried out at "Capitão de Campos" Farm, in the municipality of Piracuruca; the one by Carvalho, Teodoro and Lemos (2018), with a floristic inventory of a caatinga-cerrado ecotonal area in the north of Piauí; Pereira and Lemos (2018) in which the flora was characterized in the town of Pontal do Anel, in Luís Correia, Piauí and; Cerqueira and Lemos (2018) in which the authors performed a floristic survey on natural trails of a site with tourist potential in the north of Piauí as a subsidy to environmental education and phytodiversity conversation.

Materials and Methods

Study Area

The study was carried out in the city of Brasileira (04º09'37.2" S and 41°44'21.0" W), located 183 km from Teresina, state capital of Piauí, with the northern boundary being the Piracuruca municipality, South with Piripiri, East with São João da Fronteira and Cocal de Telha and west with Batalha, with an estimated population of 7966 inhabitants in 2010. It has an area of 881,481 km² and an altitude of 180 meters (IBGE, 2010).

The subject area of this study refers to the community "Palmeira da Emília" and surroundings, located in the city of Brasileira, near the Sete Cidades National Park. Figure 1 shows the study area with emphasis on the area where the plots for the phytosociological survey were installed. In the immediate vicinity of the place of study there are approximately eight families living in this region. Although anthropic actions occur, the study area has the flora still relatively preserved when compared to the general environment, showing a certain floristic diversity, when compared to preserved areas.

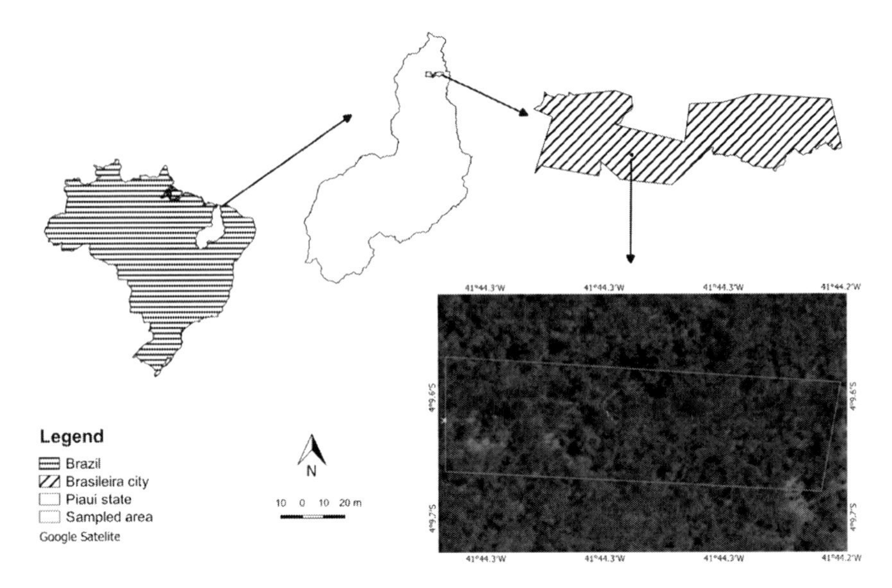

Source: Adaptated by authors.

Figure 1. Location of the municipality Brasileira, Piauí state, with the area where the plots for the phytosociological survey were installed.

Floristic Survey and Economic Potential of Plant Species

Field trips occur every fortnightly, from October 2017 to July 2018, following routine field procedures (Lawrence, 1973; Mori et al., 1989; Vaz et al., 1992). The collection of botanical material follows the method of random walks carried out to the fullest possible extent of the studied area. The collected specimens are pressed with their respective field records and herborized according to the usual Botany methodology (Silva et al., 1989).

As the collections were carried out, the species were scientifically identified and analyzed and evaluated for their usefulness (timber, medicinal, forage, oil producers, food, honey producers, ornamental, etc.), based on the specialized literature (Lewington, 1990; Rizzini and Mors, 1995; Simpson and Ogorzaly, 1995; Agra, 1996; Costa et al., 2002; Lorenzi et al., 2003; Maia, 2004) and even with the help of a person who knows the usefulness of the local plant species.

The organization of botanical families followed the Angiosperm Phylogeny Group IV system (APG IV, 2016). The botanical synonyms were updated through queries to the Brazilian Flora List of 2018 (http://floradobrasil.jbrj.gov.br), as well as the spelling of the species authors. According to the identification of the species confirmed, these were having their geographical distribution patterns verified, being done by researches in the collections of several Virtual Herbariums and the Herbarium "HDelta" (Herbarium Delta do Parnaíba), from Federal University of Piauí-UFPI/*Campus* Ministro Reis Velloso, as well as in the specialized literature.

Collection and Treatment of Soil and Climate Data

Fourteen soil samples were collected at two depths (0-20 cm and 20-40 cm), free of litter, which were analyzed by the Soil Analysis Center - *Campus* Prof. Cinobelina Elvas-UFPI. The following variables were analyzed: Water pH, K, Ca, Mg, H + Al, Al, P, Cu, Mn, Zn, Cation Exchange Capacity at pH 7.0 (CTC), Organic Matter, Sum of Exchangeable Bases, Index of Base Saturation, Index of Aluminum Saturation, Clay, Silt, Sand.

The data of the monthly average precipitation and minimum and maximum temperatures of a period of 22 years (1995-2017) of the conventional meteorological station of Piripiri city, Piauí state, provided by the Meteorological Database for Teaching and Research (BDMEP), were compiled and elaborated the climogram through Microsoft Excel software.

Phytosociology

In the phytosociological survey, the multiple plots method, commonly used in quantitative studies carried out in xeric vegetation in the semi-arid region of the Brazilian Northeast, was used as a sampling unit (Gomes, 1979; Santos, 1987; Fonseca, 1991; Rodal, 1992; Oliveira et al., 1997;

Araújo et al., 1998; Lemos and Rodal, 2002; Mendes 2003; Mesquita, 2007; Castro et al., 2009; Matos and Felfili, 2010; Lima et al., 2013).

Seven plots with a dimension of 20 x 50 m were installed semi-permanently, being distributed along a transection allocated with the aid of a compass and a 50 m track and marked with the aid of GPS (Global Positioning System). Each plot was delimited by four wooden stakes of 1m with the apex tied a red ribbon and with a string of vegetable fiber tied at an average height of 0.5 m from the ground. Within each plot, the stems of all living individuals still standing were measured, which met the criterion of diameter of 5 cm to 1.30 m of soil height. Plants that touched, inside or out, the boundary line on two sides of the plot and scattered on the other two sides were measured. Each sample was given a growing number, regardless of plot, printed on a plastic tag that was attached to individuals with string. From each individual, the following data were recorded in field records, per plot: date of observation; number of the individual; common name and/or scientific name; circumference (in centimeters) and height (in meters). The circumference measurements were performed with a metric line. For the measures of height of the individuals an adjustable rod of 6m was used (pipes of PVC insertible to each 1,5 m), placed close to the individual. The quantitative data were calculated through the FITOPAC 2.1 Program, developed by the Department of Botany of the State University of Campinas-UNICAMP (Shepherd, 2012). The following phytosociological parameters were calculated: density, frequency and relative density, frequency and absolute density, values of importance and coverage, Shannon diversity index, Equability, mean and maximum heights and diameters.

Results and Discussion

Floristic Survey and Economic Potential of the Species

The diversity of the flora in "Palmeira da Emília," Brasileira city, Piauí state, was represented by tree, shrub and herbaceous species, distributed in

55 families, 126 genera and 141 species (Table 1). The families with the greatest wealth of genera were Fabaceae (20 genera), Lamiaceae (seven genera), Rubiaceae and Malvaceae (six genera each) and Apocynaceae and Asteraceae (five genus each), Bignoniaceae (four genera). Approximately 52.7% (29) of the families were represented by a single gender at this location. The genera with the highest species richness were *Aspidosperma* (Apocynaceae) and *Mimosa* (Fabaceae), with three species each; followed by *Annona* (Annonaceae), *Bauhinia* and *Sida* (Fabaceae), *Byrsonima* (Malpighiaceae), *Campomanesia* and *Myrcia* (Myrtaceae), *Croton* (Euphorbiaceae), *Cuphea* (Lythraceae), *Fridericia* and *Handroanthus* (Bignoniaceae), *Merremia* (Convolvulaceae) *Qualea* (Vochysiaceae) and *Sida* (Malvaceae), all with two species each. It can be observed that 88% of the genera and 47.2% of the families contribute with only one species, which shows a high variation (heterogeneity) of genera and families in the studied plant community.

It was also observed the significant presence of botanical families that contributed with a species in the surveys carried out by Sousa et al., (2008), in an area of Cerrado in Maranhão; Silva and Lemos (2018) and Pereira and Lemos (2018) in transitional areas between two or more plant formations in Piauí.

The families with the highest number of species were Fabaceae (24 spp.), Lamiaceae and Malvaceae (seven spp. each), Apocynaceae, Bignoniaceae, and Rubiaceae (six spp. each), Asteraceae and Myrtaceae (five spp.), Convolvulaceae and Malpighiaceae (four spp.). These ten families accounted for 53% of the total number of species. In addition, five families were represented with three species each (10.4%), 13 families appeared with two species each (23.6%) and 27 families had one species each (49%) (Figure 2).

The Fabaceae family, which presented the highest number of species, was found in other studies developed in different plant formations present in Piauí, such as Lindoso et al. (2010) and Moura et al. (2010) in the Cerrado; Farias (2003), Castro and Mendes (2010), Amaral and Lemos (2015) and Carvalho, Teodoro and Lemos (2018) in ecotonal areas.

Table 1. List of families and species recorded in "Palmeira da Emília", Brasileira city, Northern Piauí, with their respective vulgar names, habit and Collector Number (CN) of Lucas Santos Araújo and its occurrence in others surveys of Piauí. CAA-"Caatinga" vegetation

Family/Species		Common Name	Habit	CN	Reference code				
					CAA	CER	CAR	RES	TRA
1	ACANTHACEAE								
1	*Elytraria imbricata* (Vahl) Pers.	-	Herb	116	-	-	-	-	-
2	ACHARIACEAE								
2	*Lindackeria ovata* (Benth.) Gilg	Orelha de onça	Shrub	238	-	-	6	-	-
3	AMARANTHACEAE								
3	*Alternanthera tenella* Colla	-	Sub-bush	242	-	-	-	8	15
4	*Froelichia humboldtiana* (Roem. & Schult.) Seub.	-	Herb	216	-	-	-	8,9	12
5	*Gomphrena leucocephala* Mart.	-	Sub-bush	204	-	-	-	-	-
4	AMARYLLIDACEAE								
6	*Habranthus sylvaticus* Herb.	Flor de trovão	Herb	45	2	-	-	-	-
5	ANACARDIACEAE								
7	*Anacardium occidentale* L.	Cajuí	Tree	194	-	4,5	6	8,9	11, 12,13,16,17
8	*Astronium fraxinifolium* Schott	Gonçalo-alves	Tree	195	1	4,5	-	9	12,17
9	*Myracrodruon urundeuva* Allemão	Aroeira	Tree	19	1,2	5	-	-	12,13,15,16
6	ANNONACEAE								
10	*Annona coriacea* Mart.	Araticum	Tree	154	-	4,5	-	-	11
11	*Annona leptopetala* (R.E.Fr.) H.Rainer	Ata brava	Tree	152	-	-	-	-	-

Family/Species		Common Name	Habit	CN	Reference code				
					CAA	CER	CAR	RES	TRA
12	*Ephedranthus pisocarpus* R.E.Fr.	Conduru	Tree	136	-	4	-	-	12,13
7	APOCYNACEAE								
13	*Aspidosperma multiflorum* A. DC.	Piquiá	Tree	260	2	5	-	-	12,13,17
14	*Aspidosperma pyrifolium* Mart. & Zucc.	Pereira	Tree	47	1,2.3	4,5	-	9	12,13,15.16
15	*Himatanthus drasticus* (Mart.) Plumel	Janaguba	Tree	27	-	-	-	-	-
16	*Schubertia grandiflora* Mart	-	Climb	99				8,9	
17	*Secondatia densiflora* A.DC.	Cipó canoinha	Climb	35	-	-	-	-	12
18	*Tabernaemontana catharinensis* A.DC.	-	Shrub	6/161	-	-	-	-	17
8	ARECACEAE								
19	*Astrocaryum campestre* Mart	Tucum	Palm tree	189	-	-	-	-	-
9	ASTERACEAE								
20	*Ageratum conyzoides* L.	-	Herb	127	-	-	6	-	-
21	*Eclipta prostrata* (L.) L.	-	Herb	156	-	-	-	-	-
22	*Lepidaploa remotiflora* (Rich.) H.Rob.	-	Herb	101	-	-	-	-	12
23	*Elephantopus mollis* Kunth.	-	Herb	251	-	-	-	-	-
24	*Melanthera latifolia* (Gardner) Cabrera	-	Herb	236	-	-	-	-	-
10	BIGNONIACEAE								
25	Fridericia dispar (Bureau ex K. Schum.) L. G .Lohmann								
26	Fridericia subverticillata (Bureau & K. Schum.) L. G. Lohmann								
27	*Handroanthus impetiginosus* (Mart. ex DC.) Mattos	Pau d'arco roxo	Tree	135	1	-	7	9	13,14,15,16
28	*Handroanthus serratifolius (Vahl) S. Grose*	Pau d'arco amarelo	Tree	17	1	4,5	7	-	13

Table 1. (Continued)

Family/Species		Common Name	Habit	CN	Reference code				
					CAA	CER	CAR	RES	TRA
29	*Neojobertia candolleana (Mart. ex DC.) Bureau & K. Schum.*	Cipó amarelo	Climb	5	-	-	-	10	-
		-	Climb	169	-	-	-	-	-
		-	Climb	124	-	-	-	-	-
30	*Pleonotoma castelnaei* (Bureau) Sandwith	-	Climb	34	-	-	-	-	17
11	BIXACEAE								
31	*Cochlospermum vitifolium* (Willd.) Spreng.	Algodão brabo	Tree	117	2	-	-	-	17
12	BORAGINACEAE								
32	*Cordia rufescens* A.DC	-	Shrub	30	1	5	-	-	17
33	*Heliotropium indicum* L.	-	Herb	193	-	-	6	-	15
13	BROMELIACEAE								
34	*Bromelia karatas* L.	Croatá	Herb	261	-	-	-	9	-
35	*Encholirium* sp.	Macambira	Herb	255	-	-	-	-	-
14	CACTACEAE								
36	*Pilosocereus* sp.	Xiquexique	Shrub	262	-	-	-	-	-
37	*Cereus jamacaru* DC.	Mandacaru	Tree	263	2	5	6	8,9,10	11,12,13,16
16	CARYOCARACEAE								
38	*Caryocar coriaceum* Wittm.	Pequi	Tree	25	-	4,5	6	-	-
17	CLEOMACEAE								
39	*Physostemon guianense* (Aubl.) Malme	-	Herb	210	-	-	-	-	-

Family/Species		Common Name	Habit	CN	Reference code				
					CAA	CER	CAR	RES	TRA
40	*Tarenaya spinosa* (Jacq.) Raf.	-	Shrub	192/158	-	-	-	-	-
	Tarenaya diffusa (Banks ex DC.) Soares Neto & Roalson		Herb	232	-	-	-	-	-
18	COMBRETACEAE								
41	*Combretum leprosum* Mart	Mufumbo	Shrub	123	2, 3	4,5	7	8,10	11,12,13,14,15,16,17
42	*Terminalia fagifolia* Mart.	Cascudo	Tree	180/198	-	-	6	-	12
19	CONVOLVULACEAE								
43	*Evolvulus pterocaulon Moric.*	-	Herb	61	-	-	-	-	-
44	*Ipomoea asarifolia* (Desr.) Roem. & Schult.	-	Climb	250	-	-	-	8,9,10	15
45	*Merremia aegyptia* (L.) Urb.	-	Climb	121	-	-	-	8	-
46	*Merremia cissoides (Lam.)* Hallier f.	-	Climb	252	-	-	-	-	-
20	CYPERACEAE								
47	*Cyperus surinamensis* Rottb.	-	Herb	131	-	-	-	-	-
48	*Cyperus obtusatus* (J.Presl & C.Presl) Mattf. & Kük. .	-	Herb	132/153	-	-	-	-	-
21	DILLENIACEAE								
49	*Curatella americana* L.	Sambaiba ou lixeira	Tree	13	3	4,5	-	9	13,14
50	*Davilla rugosa* Poir.	Sambaibinha	Climb	122	-	-	-	-	17
22	EUPHORBIACEAE								
51	*Cnidoscolus* ***urens*** (L.) Arthur	Urtiga	Shrub	197	1	-	-	8,9	-
52	*Croton nepetifolius* Baill	Marmeleiro	Shrub	58	-	-	-	-	-

Table 1. (Continued)

Family/Species		Common Name	Habit	CN	Reference code				
					CAA	CER	CAR	RES	TRA
53	*Croton sonderianus* Müll.Arg	Marmeleiro	Shrub	202	1	-	6	-	16,17
23	FABACEAE								
54	*Anadenanthera colubrina* (Vell.) Brenan	Angico branco	Tree	43	-	5	6	-	13,16
55	*Bauhinia ungulata* L.	Mororó	Tree	247	2	5	6,7	9	12,14,15,16
56	*Bauhinia* sp.	Mororó	Tree	60					
57	*Calliandra fernandesii* Barneby	-	Shrub	28	-	4,5	-	9	-
58	*Centrosema brasilianum* (L.) Benth.	-	Climb	248	-	-	-	8,9	15,17
59	*Chloroleucon acacioides* (Benth.) G. P. Lewis	-	Tree	159	-	4	-	8,9	-
60	*Copaifera langsdorffii* Desf.	Podoí	Tree	196	3	-	-	9	15
61	*Dahlstedtia araripensis* (Benth.) M. J. Silva & A. M. G. Azevedo	-	Tree	16	2	-	-	-	-
62	*Desmodium glabrum* (Mill.) DC.	-	Herb	269	-	-	-	8,9	-
63	*Dioclea grandiflora* Mart. ex Benth.	Mucunã	Climb	199/111	1,2	-	-	-	17
64	*Enterolobium contortisiliquum*	Tamboril	Tree	264	-	-	-	-	-
65	*Hymenaea courbaril* L.	Jatobá	Tree	44/164	1	4	-	9	12,13,14,15,16
66	*Macroptilium lathyroides* (L.) Urb.	-	Herb	139	-	-	-	-	-
67	*Mimosa caesalpiniifolia* Benth.	Sabiá	Tree	134	-	5	7	9	12,13,14,15,16,17
68	*Mimosa velloziana* Mart.	Malicia	Sub-bush	98	-	-	-	-	-

Family/Species		Common Name	Habit	CN	Reference code				
					CAA	CER	CAR	RES	TRA
69	*Mimosa verrucosa* Benth.	-	Shrub	110	1	4,5	-	9	16
70	*Parkia platycephala* (Willd.) Benth. ex Walp.	Faveira	Tree	190	-	4,5	-	9	16
71	*Periandra coccinea* (Schrad.) Benth.	-	Climb	115	-	-	-	-	-
72	*Pityrocarpa moniliformis* (Benth.) Luckow & R. W. Jobson	Catanduva	Tree	212	-	-	-	9	15,16
73	*Platypodium elegans* Vogel	-	Tree	267	1	-	-	9	-
74	*Poincianella gardneriana* (Benth.) L. P. Queiroz	Catingueira	Shrub	33	-	-	-	-	17
75	*Senna alata* (L.) Roxb.	-	Tree	249	-	-	-	9	-
76	*Senna* sp.	Besouro	Shrub	3	-	-	-	-	-
77	*Tachigali vulgaris* L. G. Silva & H. C. Lima	Pau pombo	Tree	26	-	-	-	-	-
24	KRAMERIACEAE								
78	*Krameria tomentosa* A. St.-Hil.	-	Shrub	56	1	5	-	9	-
25	LAMIACEAE								
79	*Amasonia campestris* (Aubl.) Moldenke	-	Sub-bush	213	2	-	6	8,9,10	12,15,17
80	*Hypenia salzmannii* (Benth.) Harley	-	Herb	188	-	-	-	-	-
81	*Hyptis atrorubens* Poit.	-	Herb	168	-	-	7	-	12
82	*Marsypianthes chamaedrys (Vahl) Kuntze*	-	Herb	206	-	-	-	8,9	17
83	*Mesosphaerum suaveolens (L.) Kuntze*	-	Sub-bush	259	-	-	-	9	15
84	*Rhaphiodon echinus Schauer*	-	Herb	109	-	-	-	-	-

Table 1. (Continued)

Family/Species		Common Name	Habit	CN	Reference code				
					CAA	CER	CAR	RES	TRA
85	*Vitex flavens* Kunth	Mama cachorra ou taturumã	Tree	7	-	4	-	-	-
26	LAURACEAE								
86	*Cassytha filiformis* L.	-	Climb	256	-	-	-	8,9,10	-
27	LECYTHIDACEAE								
87	*Lecythis pisonis* Cambess.	Sapucaia	Tree	257	-	-	-	9	-
28	LORANTHACEAE								
88	*Psittacanthus robustus* (Mart.) Mart.	Enxerte	Herb	31/173	-	-	-	8,9	-
29	LYTHRACEAE								
89	*Cuphea ericoides* Cham. & Schltdl	-	Shrub	243	2	-	-	-	-
90	*Cuphea melvilla* Lindl.	-	Shrub	113	-	-	-	-	-
30	MALPIGHIACEAE								
91	*Byrsonima correifolia* A.Juss.	Murici	Shrub	112	2	4,5	-	-	13,16
92	*Byrsonima crassifolia* (L.) Kunth	Murici	Tree	14	-	4,5	-	9	13,17
93	*Diplopterys pubipetala* (A.Juss.) W.R.Anderson & C.C.Davis	-	Climb	162	-	-	-	-	-
94	*Heteropterys eglandulosa* A.Juss.	-	Climb	41	-	-	-	-	-
31	MALVACEAE								
95	*Guazuma ulmifolia* Lam.	Mutamba	Tree	9	-	-	-	-	-
96	*Helicteres heptandra L.B.Sm.*	Caçatrapi	Shrub	54/138	-	-	-	-	-
97	*Melochia tomentosa L.*	-	Shrub	222	1	-	-	-	-

Family/Species		Common Name	Habit	CN	Reference code				
					CAA	CER	CAR	RES	TRA
98	*Pavonia cancellata (L.) Cav.*	-	Herb	94	-	-	-	-	17
99	*Sida cordifolia* L.	-	Sub-bush	217	1	-	-	-	-
100	*Sida rhombifolia* L.	Relógio	Herb	227	-	-	-	-	-
101	*Waltheria indica* L.	-	Herb	253	-	-	-	8,10	12,17
32	MELASTOMATACEAE								
102	*Mouriri guianensis* Aubl.	Criuli	Tree	29	-	-	-	-	-
33	MOLLUGINACEAE								
103	*Mollugo verticillata* L.	-	Herb	234	1	-	-	9,10	-
34	MYRTACEAE								
104	*Campomanesia aromatica* (Aubl.) Griseb.	Guabiraba amarela	Tree	160	-	5	7	10	16,17
105	*Campomanesia velutina* (Cambess.) O. Berg	-	Tree	176	-	-	-	-	-
106	*Eugenia flavescens* DC.	-	Tree	182	2	-	-	-	-
107	*Myrcia guianensis* (Aubl.) DC.	Farinha seca	Tree	148	-	5	-	8	-
108	*Myrcia multiflora* (Lam.) DC.	Maria preta	Tree	18	-	4	-	8,9	17
35	OCHNACEAE								
109	*Ouratea hexasperma* (A. St.-Hil.) Baill.	-	Tree	165	-	5	-	-	17
36	OLACACEAE								
110	*Ximenia americana* L.	-	Tree	8	1,2,3	5	-	9,10	12,16,17
37	ONAGRACEAE								
111	*Ludwigia leptocarpa* (Nutt.) H. Hara	-	Shrub	266	-	-	-	-	-
38	OPILIACEAE								
112	*Agonandra brasiliensis* Miers ex Benth. & Hook.f.	Marfim	Tree	265	-	-	-	-	-

Table 1. (Continued)

Family/Species		Common Name	Habit	CN	Reference code				
					CAA	CER	CAR	RES	TRA
39	OXALIDACEAE								
113	*Oxalis divaricata* Mart. ex Zucc.	-	Herb	214	-	-	-	-	17
40	PASSIFLORACEAE								
114	*Passiflira* sp.	-	Climb	219	-	-	-	-	-
41	PHYLLANTHACEAE								
115	*Phyllanthus niruri* L.	Quebra pedra	Herb	268	-	-	-	-	16
42	PLANTAGINACEAE								
116	*Angelonia cornigera* Hook.f.	-	Herb	125	-	-	-	-	-
117	*Scoparia dulcis* L.	-	Herb	254	1	-	-	9	15,16
43	POACEAE								
118	*Streptostachys asperifolia* Desv.	-	Herb	218	-	-	-	8,9	-
44	POLYGALACEAE								
119	*Asemeia violacea* (Aubl.) J. F. B. Pastore & J. R. Abbott	-	Herb	207	-	-	-	9	-
120	*Polygala paniculata* L.	-	Herb	130	-	-	-	-	-
45	RUBIACEAE								
121	*Chomelia anisomeris* Müll.Arg.	-	Shrub	53	-	-	-	-	-
122	*Cordiera sessilis* (Vell.) Kuntze	-	Shrub	15	-	-	-	-	-
123	*Genipa americana* L.	Jenipapo	Tree	1	-	-	-	9	-
124	*Guettarda viburnoides* Cham. & Schltdl.	Angeuca	Tree	64	-	-	-	-	17
125	*Mitracarpus frigidus* (Willd. ex Roem. & Schult.) K.Schum.	-	Sub-bush	100	-	-	-	-	-

Family/Species		Common Name	Habit	CN	Reference code				
					CAA	CER	CAR	RES	TRA
126	*Richardia grandiflora* (Cham. & Schltdl.) Steud.	-	Herb	258	-	-	-	8,9	-
46	SALICACEAE								
127	*Casearia grandiflora* Cambess	-	Shrub	120	1	4	6	-	-
47	SANTALACEAE								
128	*Phoradendron quadrangulare* (Kunth) Griseb.	-	Herb	184	2	-	-	-	-
48	SAPINDACEAE								
129	*Magonia pubescens* A. St.-Hil.	Tingui	Tree	187	2	5	-	9	12,13
130	*Paullinia pinnata* L.	-	Climb	36	-	-	-	-	-
131	*Talisia esculenta* (Cambess.) Radlk.	-	Tree	20	1	-	7	-	-
49	SAPOTACEAE								
132	*Pouteria ramiflora* (Mart.) Radlk.	Pitomba de leite	Tree	117	-	-	-	-	16
50	SIMAROUBACEAE								
133	*Homalolepis cedron* (Planch.) Devecchi & Pirani	Canjarana	Tree	11	-	-	-	-	17
134	*Simarouba versicolor* A. St.-Hil.	Paraíba	Tree	12	3	4,5	-	9	12,13,16,17
51	SOLANACEAE								
135	*Solanum paniculatum* L.	Jurubeba	Shrub	21	-	-	-	-	-
52	VERBENACEAE								
136	*Lantana camara* L	-	Shrub	221	1	-	6	8,9	12,14
137	*Lippia grata* Schauer	Alecrim	Shrub	108	-	-	-	-	-
53	VIOLACEAE								

Table 1. (Continued)

Family/Species		Common Name	Habit	CN	Reference code				
					CAA	CER	CAR	RES	TRA
138	*Pombalia calceolaria (L.)* Paula-Souza	-	Herb	205	-	-	-	9	-
54	VITACEAE								
139	*Cissus erosa* Rich.	-	Climb	244	-	-	-	8,9	-
55	VOCHYSIACEAE								
140	*Qualea grandiflora* Mart.	Pau Terra	Tree	224	3	4,5	-	9	-
141	*Qualea parviflora* Mart.	-	Tree	48	1	4,5	-	9	12,13

1-Lemos, 2004; 2- Mendes and Castro, 2010; 3-Alves et al., 2013), CER-"Cerrado" vegetation (4-Mesquita and Castro, 2007; 5-Sousa et al., 2009) CAR-"Carrasco" vegetation (6-Chaves et al., 2005; 7-Chaves et al., 2007) RES- "Restinga" vegetation (8-Santos-Filho, 2009; 9-Santos-Filho et al., 2015; 10-Santos-Filho et al., 2016), TRA- Transition vegetation (11-Oliveira et al., 1997; 12-Farias, 2003; 13-Barros, 2005; 14-Amaral et al., 2012; 15-Amaral and Lemos, 2015; 16-Sousa et al., 2015; 17-Carvalho, Teodoro and Lemos, 2018

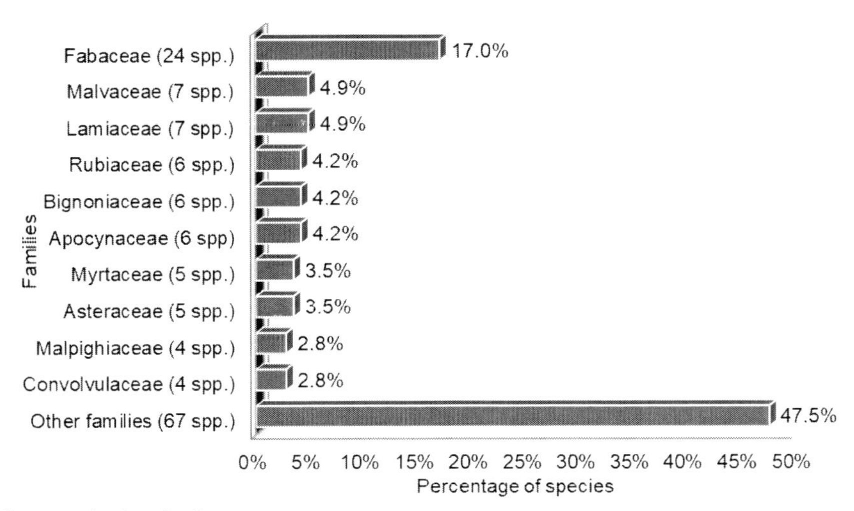

Source: Authors's data.

Figure 2. Families with the highest representativeness, in percentage of number of species, in the floristic survey in "Palmeira da Emília", Brasileira city, North of Piauí.

Fabaceae accounted for 17% of the number of species, agreeing with data from the Brazil Flora Group (BFG, 2015) as the largest family in almost all of the country's ecosystems and biomes. In this regard, this family is also among the most representative in surveys carried out in several vegetation formations in Piauí, such as those of Lemos (2004), in the "caatinga"; Mesquita and Castro (2007), in the "cerrado"; Castro et al. (2009), in Seasonal Forest; Matos et al. (2010), in "Matas de Galerias"; Oliveira et al. (2007), Silva and Lemos (2018), Pereira and Lemos (2018) in transition areas and; Santos-Filho et al. (2015) in "restinga."

Lamiaceae accounted for 4.9% of the number of species. Being a cosmopolitan family, it presents genera of economical and medicinal importance, like some species of the genera *Melissa*, *Mentha*, *Origanum*, *Rosmarinus*, *Salvia*, *Salvia*, *Satureja* and *Thymus*, used mainly in the cooking in certain regions of the Mediterranean and Lavandula and Pogostemon in the manufacture of perfumes for being rich in essential oils (Harley et al., 2004).

Apocynaceae accounting for 4.9% of the number of species, is a family with wide distribution, occurring in tropical, subtropical and certain genera occur in temperate areas. It is among the ten largest families of angiosperms in the world and in Brazil, only the Pampa and Pantanal are not among the ten main families of angiosperms per biome. Certain species have great commercial, logging, ornamental and medicinal utility (BFG, 2015; Rapini, 2000). Apocynaceae is also represented in vegetational formations in Piauí, such as Santos-Filho (2009) in Restinga and Sousa, Araújo and Lemos (2015) in a Cerrado/Caatinga transition area in the municipality of Buriti dos Lopes, north of Piauí.

Bignoniaceae represented 4.1% of the number of species, being a family that occupies the sixth place of representation in the Brazilian biome "Pantanal," according to Brazil Flora Group (BFG, 2015), but also has a good representation in studies carried out in Piauí, such as Oliveira et al. (1997), in Carrasco-caatinga transition; Moura et al. (2010), in Cerrado and Castro et al. (2014) in Transitional Seasonal Forest.

So far, only *Myracrodruon urundeuva* Allemão was found included in the official list of endangered species in the "vulnerable" category of the Brazilian Institute of Environment and Renewable Natural Resources-IBAMA (www.ibama.gov.br). No other rare species were found in the studied area, according to the "Red Book of Flora of Brazil" (MARTINELLI, 2013) and "Rare Plants of Brazil" (Giulietti et al., 2009). Vieira-Filho, Meireles and Lemos (2018) reported these species in a rural community in Piauí, as being the most difficult species to be found in the community, due to the uncontrolled exploitation for construction purposes.

From the 141 species, plants of arboreal habit represent 38.2% (52 spp.), and this habit occurs more frequently in the Amazonian Forest and Atlantic Forest in Brazil (BFG, 2015). In this study, most species represented by the tree stratum belong to the family Fabaceae, such as *Anadenanthera colubrina* (Vell.) Brenan, *Bauhinia ungulata* L., *Cassia alata*, *Chloroleucon acacioides* (Benth.) GPLewis, *Copaifera langsdorffii* Desf., *Dahlstedtia araripensis* (Benth.) MJ Silva & AMG Azevedo, *Enterolobium contortisiliquum* (Vell.) Morong., *Hymenaea courbaril* L., *Mimosa caesalpiniifolia* Benth., *Parkia platycephala* (Willd.) Benth. ex Walp., *Pityrocarpa moniliformis* (Benth.) Luckow & R. W. Jobson and *Platypodium elegans* Vogel, *Tachigali vulgaris* L. G. Silva & H. C. Lima and; Myrtaceae with representatives such as *Campomanesia aromatica* (Aubl.) Griseb., *Campomanesia velutina* (Cambess.) O. Berg, *Eugenia flavescens* DC., *Myrcia guianensis* (Aubl.) DC. and *Myrcia multiflora* (Lam.) DC.

In addition to the families mentioned above, the tree stratum was also well represented by the families Anacardiaceae (*Anacardium occidentale* L., *Astronium fraxinifolium* Schott, *Myracrodruon urundeuva* Allemão), Annonaceae (*Annona coriacea* Mart, *Annona leptopetala* (R.E.Fr.) H. Rainer, *Ephedranthus pisocarpus* R.E.Fr.) and Apocynaceae (*Aspidosperma multiflorum* A. DC., *Aspidosperma pyrifolium* Mart. & Zucc., *Himatanthus drasticus* (Mart.) Plumel).

The herbaceous stratum is represented by 25.7% of the total species (34 spp.). Its composition for the Cerrado, for example, presents an important ecological role, contributing to the maintenance of the woody flora of this place, considering that there is a great representation in the Cerrado and Atlantic Forest (BFG, 2015). The soils covered by herbs are preserved from erosion and sustain the high temperature and humidity, forming a mechanism of hot and humid natural germination. The roots of the herbs shuffle in the superficial portion of the soil, creating a web that assists in the containment of seeds of both tree and herbaceous plants, favoring the renewal of plant populations (Ferraz, 2009).

The species that represented this life form in the study area are *Cyperus surinamensis* Rottb., *Elephantopus mollis* Kunth., *Habranthus sylvaticus* Herb., *Heliotropium indicum* L., *Hypenia salzmannii* (Benth.) Harley, *Hyptis atrorubens* Poit., *Kyllinga vaginata* Lam., *Macroptilium lathyroides* (L.) Urb., *Pavonia cancellata* (L.) Cav. and *Rhaphiodon echinus* Schauer.

The proportion of shrub life in this survey is 18.3% (26 spp.), occurring more frequently in Atlantic Forest and Cerrado (BFG, 2015). In the studied area, the species belonging to the shrub stratum do not belong expressively to a single family. The species with this life form were *Byrsonima correifolia* A.Juss., *Cnidoscolus urens* (L.) Arthur, *Combretum leprosum* Mart, *Cordia rufescens* A.DC., *Cordiera sessilis* (Vell.) Kuntze, *Helicteres heptandra* L.B.Sm., *Lippia grata* Schauer., *Poincianella gardneriana* (Benth.) L.P.Queiroz, *Solanum paniculatum* L. and *Tarenaya spinosa* (Jacq.) Raf.

Lianas are represented by 12.5% (19 spp.) In this survey and, among Brazilian biomes, are frequently recorded in the Atlantic Forest, Amazon Forest and Cerrado (BFG, 2015). The lianas help to stabilize the microclimate of the forests and, thus, enhance the conditions for germination and seedling formation of primary tree species. The leaves of the lianas help to stabilize the microclimate in the cold and dry season, when most of the tree and shrub composition lose the leaves (Engel, Fonseca and Oliveira, 1998).

The species of lianas that were observed in the study area were *Davilla rugosa* Poir., *Dioclea grandiflora* Mart. ex Benth., *Diplopterys pubipetala* (A. Juss.) W. R. Anderson & C. C. Davis, *Fridericia dispar* (Bureau ex K. Schum.) L. G. Lohmann., *Fridericia subverticillata* (Bureau & K. Schum.) L. G. Lohmann., *Merremia aegyptia* (L.) Urb., *Momordica charantia* L., *Neojobertia candolleana* (Mart. ex DC.) Bureau & K. Schum., *Paullinia pinnata* L., *Periandra coccinea* (Schrad.) Benth. and *Secondatia densiflora* A.DC.

The percentage of habit of the species recorded in the studied area can be observed in Figure 3.

In this study, 35 species (25.3%) endemic to Brazil were registered, according to "Flora do Brasil 2020", being a high percentage, considering that according to the Brazil Flora Group (BFG, 2015), there was a sharp fall in the number of endemic species (-9.1%) in 2010, with the occurrence of a 40.7% increase in species for the State, from 1,416 in 2010 to 1,992 species in 2015.

The endemic species recorded were *Angelonia cornigera* Hook.f., *Annona leptopetala* (R.E.Fr.) H. Rainer, *Aspidosperma multiflorum* A. DC., *Byrsonima correifolia* A. Juss., *Calliandra fernandesii* Barneby, *Campomanesia velutina* (Cambess.) O. Berg, *Caryocar coriaceum* Wittm., *Cereus jamacaru* DC., *Chomelia anisomeris* Müll.Arg., *Cissus erosa* Rich., *Cordia rufescens* A.DC., *Croton sonderianus* Müll.Arg, *Cuphea ericoides* Cham. & Schltdl, *Dioclea grandiflora* Mart. ex Benth., *Ephedranthus pisocarpus* R.E.Fr., *Fridericia dispar* (Bureau ex K.

Schum.) L. G. Lohmann, *Fridericia subverticillata* (Bureau & K. Schum.) L. G. Lohmann, *Gomphrena leucocephala* Mart., *Habranthus sylvaticus* Herb., *Lecythis pisonis* Cambess., *Lindackeria ovata* (Benth.) Gilg, *Mimosa caesalpiniifolia* Benth., *Mimosa verrucosa* Benth, *Neojobertia candolleana* (Mart. ex DC.) Bureau & K. Schum., *Oxalis divaricata* Mart. ex Zucc., *Periandra coccinea* (Schrad.) Benth., *Pityrocarpa moniliformis* (Benth.) Luckow & R. W. Jobson, *Pleonotoma castelnaei* (Bureau) Sandwith, *Poincianella gardneriana* (Benth.) L.P.Queiroz, *Rhaphiodon echinus* Schauer, *Tachigali vulgaris* L.G.Silva & H.C.Lima, *Talisia esculenta* (Cambess.) Radlk e *Tarenaya diffusa* (Banks ex DC.) Soares Neto & Roalson.

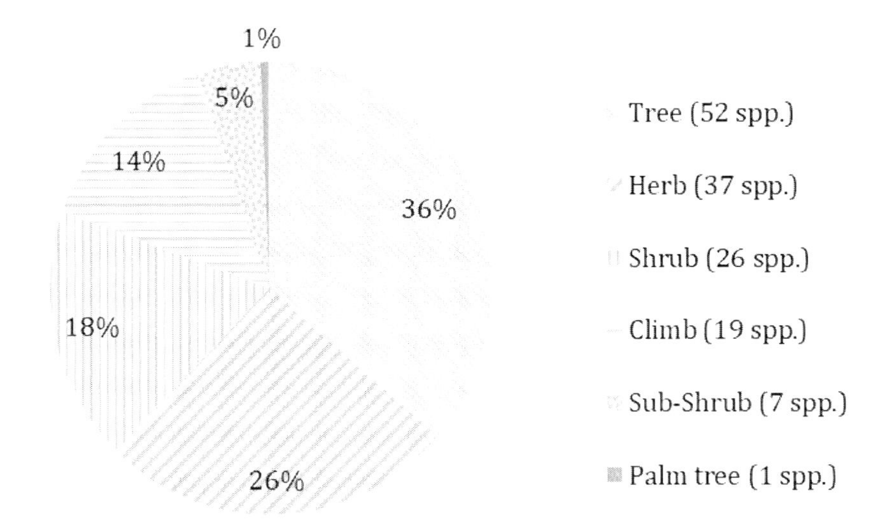

Source: Authors's data.

Figure 3. Distribution of the species by habit in the floristic survey in "Palmeira da Emília", Brasileira city, north of Piauí.

From the collected and identified plants, 57 species distributed in 32 families have economic potential, according to the specialized literature (Table 2).

Table 2. List of families and species registered with economic potential in "Palmeira da Emília," Brasileira city, north of Piauí, with their respective common names, habit and use categories: A = Medicinal, B = Forage, C = Timber, D = Food, E = Ornamental, F = Fuel, G = Honey producer

Family/Species		Common Name	Habit	Use Categories
1	AMARANTHACEAE			
1	*Alternanthera tenella* Colla	-	Sub-bush	B
2	*Froelichia humboldtiana* (Roem. & Schult.) Seub.	-	Herb	B
2	AMARYLLIDACEAE			
3	*Habranthus sylvaticus* Herb.	Flor de trovão	Herb	A,E
3	ANACARDIACEAE			
4	*Anacardium occidentale* L.	Cajuí	Tree	A,B,D,E,G
5	*Astronium fraxinifolium* Schott	Gonçalo-alves	Tree	C,E,G
6	*Myracrodruon urundeuva* Allemão	Aroeira	Tree	A,C,E,G
4	ANNONACEAE			
7	*Annona coriacea* Mart.	Araticum	Tree	A,B,D,G
5	APOCYNACEAE			
8	*Aspidosperma pyrifolium* Mart. & Zucc.	Pereira	Tree	C
9	*Aspidosperma multiflorum* A. DC.	Piquiá	Tree	C, F
6	BIGNONIACEAE			
10	*Handroanthus impetiginosus* (Mart. ex DC.) Mattos	Pau d'arco roxo	Tree	A,B,C,E,F,G
11	*Handroanthus serratifolius* (Vahl) S.Grose	Pau d'arco amarelo	Tree	A,C,E,F,G
12	*Neojobertia candolleana* (Mart. ex DC.) Bureau & K. Schum.	Cipó amarelo	Climb	E, G
7	BIXACEAE			
13	*Cochlospermum vitifolium* (Willd.) Spreng.	Algodão brabo	Tree	B,G

Family/Species		Common Name	Habit	Use Categories
8	BORAGINACEAE			
14	*Cordia rufescens* A.DC	-	Shrub	A,B,D,G
15	*Heliotropium indicum* L.	-	Herb	F
9	BROMELIACEAE			
16	*Bromelia karatas* L.	Croatá	Herb	A
10	CACTACEAE			
17	*Cereus jamacaru* DC.	Mandacaru	Tree	D, E
11	CARYOCARACEAE			
18	*Caryocar coriaceum* Wittm.	Pequi	Tree	A,B,C,D,E,F,G
12	COMBRETACEAE			
19	*Combretum leprosum* Mart	Mufumbo	Shrub	B,A,G,E
20	*Terminalia fagifolia* Mart.	-	Tree	A, C, F, G
13	CONVOLVULACEAE			
21	*Merremia aegyptia* (L.) Urb.	-	Climb	G
22	*Ipomoea asarifolia* (Desr.) Roem. & Schult.	-	Climb	A
14	CUCURBITACEAE			
23	*Momordica charantia L.*	Melão de são Caetano	Climb	A,D
15	DILLENIACEAE			
24	*Curatella americana* L.	Sambaíba	Tree	A,C,E,F,G
16	FABACEAE			
25	*Bauhinia ungulata* L.	Mororó	Tree	A, F, G
26	*Calliandra fernandesii* Barneby	-	Shrub	E
27	*Dioclea grandiflora* Mart. ex Benth.	Mucunã	Climb	D,G
28	*Enterolobium contortisiliquum*	Tamboril	Tree	A
29	*Hymenaea courbaril* L.	Jatobá	Tree	A,C,D,F,G
30	*Mimosa caesalpiniifolia* Benth.	Sabiá	Tree	A,B,C,F,G

Table 2. (Continued)

Family/Species		Common Name	Habit	Use Categories
31	*Mimosa velloziana* Mart.	Malicia	Sub-bush	A, B, G
32	*Mimosa verrucosa* Benth.	-	Shrub	C, F
33	*Parkia platycephala* (Willd.) Benth. ex Walp.	Faveira	Tree	A,B,C,E,F,G
17	KRAMERIACEAE			
34	*Krameria tomentosa* A.St.-Hil.	-	Shrub	A, G
18	LAMIACEAE			
35	*Hypenia salzmannii* (Benth.) Harley	-	Herb	A
36	*Rhaphiodon echinus* Schauer	-	Herb	A,B
19	LYTHRACEAE			
37	*Cuphea ericoides* Cham. & Schltdl	-	Shrub	G
20	MALPIGHIACEAE			
38	*Byrsonima correifolia* A.Juss.	Murici	Shrub	A,B,C,D,E,F,G
39	*Byrsonima crassifolia* (L.) Kunth	Murici	Tree	D
21	MALVACEAE			
40	*Helicteres heptandra* L.B.Sm.	Caçatrapi	Shrub	A
41	*Pavonia cancellata* (L.) Cav.	-	Herb	A, E
42	*Sida cordifolia* L.	-	Sub-bush	A
22	MYRTACEAE			
43	*Campomanesia velutina* (Cambess.) O.Berg	-	Tree	A, B, D, G
23	OCHNACEAE			
44	*Ouratea hexasperma* (A.St.-Hil.) Baill.	-	Tree	F
24	OLACACEAE			
45	*Ximenia americana* L.	-	Tree	A,B,C,D,G

Family/Species		Common Name	Habit	Use Categories
25	OPILIACEAE			
46	*Agonandra brasiliensis* Miers ex Benth. & Hook.f.	Malfim	Tree	D
26	OXALIDACEAE			
47	*Oxalis divaricata* Mart. ex Zucc.	-	Herb	D
27	PLANTAGINACEAE			
48	*Angelonia cornigera* Hook.f.	-	Herb	E
49	*Scoparia dulcis* L.	-	Herb	A
28	RUBIACEAE			
50	*Guettarda viburnoides* Cham. & Schltdl.	-	Tree	B, G
51	*Richardia grandiflora* (Cham. & Schltdl.) Steud	-	Herb	B
29	SAPINDACEAE			
52	*Magonia pubescens* A.St.-Hil.	-	Tree	A,B,C,E,F,G
53	*Talisia esculenta* (Cambess.) Radlk.	-	Tree	D,G
30	SIMAROUBACEAE			
54	*Simarouba versicolor* A.St.-Hil.	Paraíba	Tree	A, F
31	SOLANACEAE			
55	*Solanum paniculatum* L.	Jurubeba	Shrub	A
32	VERBENACEAE			
56	*Lantana camara* L	-	Shrub	E
33	VOCHYSIACEAE			
57	*Qualea grandiflora* Mart.	Pau terra	Tree	C, A, G, F
58	*Qualea parviflora* Mart.	-	Tree	A,C,E,F,G

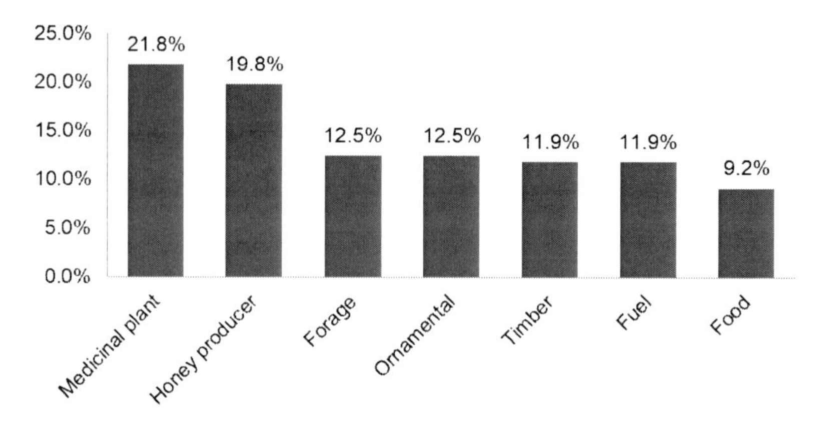

Source: Authors's data.

Figure 4. Percentage of citations by category of use in the study in "Palmeira da Emília", Brasileira, north of Piauí.

The most representative families with economic potential were Fabaceae (nine spp.) And Anacardiaceae, Bignoniaceae and Malvaceae (three spp. each). Thus, most have economic potential in the category of medicinal (35 spp.) and honey producer (30 spp.), followed by ornamental and forage (19 spp. each), fuel and timber (18 spp. each) and food (15 spp.). From the total number of plants that have economic potential, 22 were referred to only one category of use; 13 for two; three to three; seven to four; seven to five; three to six and two species to seven use categories.

The number of species (35 spp.) was highlighted in the category of medicinal use (Figure 4), in agreement with the data found by Chaves (2005) in "Carrasco" vegetation; Franco (2006) in a Cerrado/Palms vegetation transition area; Oliveira et al. (2010), Sousa, Araújo and Lemos (2015) in a transition area Cerrado/Caatinga; Vieira-Filho, Meireles and Lemos (2018) in a Caatinga-Cerrado-Restinga transition area. It is emphasized that the habit of using plants to combat routine health problems is something that is part of people's daily lives, both in rural communities and in urban centers.

The use of plants in folk medicine provides a financial saving in the purchase of medicines in compounding pharmacies. In regions where the passage to the urban center becomes more difficult because of the distance, it is noticed that there is a dependence between the user and herbal medicine and consequently the interaction with the flora, generating the transmission of empirical botanical knowledge (Silva, 2010).

The melliferous plant category was the second one with the highest number of species (30). Vilela (2000) reports that apiculture can be used by family farmers in the region from an activity planning over the years. The development of apiculture in Piauí state represents an alternative of great socioeconomic importance. Chaves, Barros and Araújo (2007), in the Middle North Basin and Chapada do Araripe, found that among the species most visited by bees are *Campomanesia aromatica* (Aubl.) Griseb., *Handroanthus impetiginosus* (Mart. ex DC.) Mattos. and *Handroanthus serratifolius* (Vahl) S. Grose, species also occurring in the studied area.

The species that had the highest number of records with several categories of use in the literature were *Byrsonima correifolia* A. Juss., *Caryocar coriaceum* Wittm. with seven categories of use each; *Handroanthus impetiginosus* (Mart. ex DC.) Mattos, *Magonia pubescens* A. St.-Hil. In addition, *Parkia platycephala* (Willd.) Benth. ex Walp., presented six types of uses each and; *Anacardium occidentale* L., *Handroanthus serratifolius* (Vahl) S. Grose, *Curatella americana* L., *Hymenaea courbaril* L., *Mimosa caesalpiniifolia* Benth., *Ximenia americana* L. and *Qualea parviflora* Mart. had five records of uses each.

Geographic Distribution of Species in Piauí State

The studied area shares species of different plant formations according to the comparisons made in different floristic surveys in Piauí, in the most diverse plant formations. It was possible to compare common species to the studied area, namely: 57 species in transition vegetation "cerrado"-"caatinga" and "caatinga"-"cerrado"; 52 species in "restinga" vegetation; 38 species in "caatinga" vegetation; 33 species in "cerrado" vegetation and 20 species in "carrasco" vegetation.

From the species recorded in the study area only 30% (42 spp.) were not mentioned in any of the surveys analyzed: *Agonandra brasiliensis*

Miers ex Benth. & Hook. f., *Angelonia cornigera* Hook. f, *Astrocaryum campestre* Mart, *Campomanesia velutina* (Cambess.) O. Berg, *Chomelia anisomeris* Müll.Arg, *Cordiera sessilis* (Vell.) Kuntze, *Cuphea melvilla* Lindl., *Cyperus surinamensis* Rottb., *Diplopterys pubipetala* (A. Juss.) W. R. Anderson & C. C. Davis, *Eclipta prostrata* (L.) L., *Elephantopus mollis* Kunth, *Elytraria imbricata* (Vahl) Pers., *Enterolobium contortisiliquum*, *Evolvulus pterocaulon* Moric., *Fridericia dispar* (Bureau ex K.Schum.) L. G. Lohmann, *Fridericia subverticillata* (Bureau & K. Schum.) L. G. Lohmann, *Gomphrena leucocephala* Mart., *Guazuma ulmifolia* Lam., *Helicteres heptandra* L. B. Sm., *Heteropterys eglandulosa* A. Juss., *Himatanthus drasticus* (Mart.) Plumel, *Hypenia salzmannii* (Benth.) Harley, *Kyllinga vaginata* Lam., *Lippia grata* Schauer, *Ludwigia leptocarpa* (Nutt.) H. Hara, *Macroptilium lathyroides* (L.) Urb., *Melanthera latifolia* (Gardner) Cabrera, *Merremia cissoides* (Lam.) Hallier f., *Mimosa velloziana* Mart., *Mitracarpus frigidus* (Willd. ex Roem. & Schult.) K. Schum., *Mouriri guianensis* Aubl, *Parkia platycephala* (Willd.) Benth. ex Walp.., *Paullinia pinnata* L., *Periandra coccinea* (Schrad.) Benth., *Physostemon guianense* (Aubl.) Malme, *Polygala paniculata* L., *Rhaphiodon echinus* Schauer, *Sida rhombifolia* L., *Solanum paniculatum* L., *Tachigali vulgaris* L. G. Silva & H. C. Lima, *Tarenaya diffusa* (Banks ex DC.) Soares Neto & Roalson and *Tarenaya spinosa* (Jacq.) Raf.

In view of these results, there is an apparent vegetation formation with own species in the studied area, since most of the occurring species (30%) are not widely distributed in the northern mesoregion of the Piauí state. This fact may be due to a still incipient number of floristic and phytosociological surveys carried out in this section of Piauí. Another fact that may contribute to this, as perceived, is that there seems to be a lack of registration of the species listed in the herbaria in the databases that foment the botanical research in Brazil and abroad.

It is noteworthy that *Combretum leprosum* Mart, *Cereus jamacaru* DC. and *Bauhinia ungulata* L. were widely distributed in all comparative plant formations. The species *C. leprosum* Mart presented a wide distribution also in the survey conducted by Silva and Lemos (2018) in a transition area of "caatinga"-"cerrado," in which the same authors commented, in fact, that this species has a wide geographical amplitude. Farias (2003) cites in ecotone areas *B. ungulata* L. among the species with the highest IVC, and Chaves (2005) in "caatinga" vegetation mentions *B. ungulata* L. among the species with the highest values of use.

Soil Data

The results of the chemical and physical analyzes of the soil samples are shown in Table 3. The exchangeable cations contents presented low values (K, Ca, Mg) and decreased with depth. The potassium saturation of the Cation Exchange Capacity (C. E. C.) at pH 7.0 at both depths did not change much, remaining at 1.1% (0-20) and 1.2% (20-40), a figure that was far below the ideal level (Sobral, 2015). The percentage of C. E. C. calcium saturation at pH 7.0 ranged from 8.8% (0-20cm) to 6.5% (20-40cm), showing a calcium imbalance and magnesium saturation, with 3.5% (0-20 and 20-40cm), according to Sobral (2015). The results of micronutrient analysis (Copper, Manganese, Zinc) were low (Sobral, 2015). As for the soil texture, the amount of sand, clay, silt did not vary with depth, but the amount of sand was much higher in relation to clay and silt, this is a characteristic of sandy soil and quartz-sand neosol, in gallery forests, a plant formation present in the Cerrado biome (Matos, 2009).

The sum of the bases (S) presented low values, which varied from 0.38 to 0.24. The C. E. C. was low (2.84 to 2.13). The values of the saturation by bases (V), were smaller than 50%, characterizing a dystrophic soil. The organic matter contents were low, varying from 3 to 5, probably due to the acidity of the soil, since the pH in water indicated values from 4.5 to 4.7.

According to CEPRO (1996), Brazilian soil is classified as a red-yellow Latosol associated with hydromorphic soils and eutrophic alluvial soils and, according to Ribeiro and Walter (1998), the *stricto sensu* Cerrado vegetation soils are of the Red-Dark Latosol, Red-Yellow Latosol and Purple Latosol Roxo, and are very or moderately acid soils (pH between 4.5 and 5.5), lacking the essential nutrients, presenting high rates of aluminum and organic matter varying from medium to low.

Table 3. Chemical and physical data of the soil in a section of vegetation of an area in "Palmeira da Emília", Brasileira, north of Piauí

Soil Variables	Depth					
	0-20	20-40	0-20		20-40	
	Average	Average	Maximum	Minimum	Maximum	Minimum
Water pH	4,5	4,7	4,7	4,4	5	4,4
K (mg/dm³)	12,84	10,74	17,6	9,3	22,2	7,4
Ca (cmol$_c$/dm³)	0,24	0,14	0,97	0,07	0,37	0,08
Mg (cmol$_c$/dm³)	0,10	0,07	0,15	0,05	0,15	0,04
H+Al (cmol$_c$/dm³)	2,45	1,88	4,03	1,78	2,27	1,67
Aluminum (cmol$_c$/dm³)	0,85	0,77	1,2	0,4	1	0,6
Phosphorus P (mg/dm³)	0,26	0,13	0,52	0	0,26	0,06
Cu	0,36	0,42	0,55	0,24	0,54	0,33
Mn	0,30	0,58	0,66	0,1	0,83	0,36
Zn	0,09	0,11	0,14	0,02	0,17	0,02
C. E. C. at pH	2,84	2,13	4,39	2,09	2,4	1,85
Organic matter (g/Kg)	5,04	3,02	10,5	3,1	7	0,7
Sum of bases (S)	0,38	0,24	1,16	0,16	0,51	0,14
Base saturation (V%)	12,97	11,44	33,9	7	23,3	5,7
Saturation Al (m%)	71,42	76,35	83,6	40,8	85,5	54,1
Clay (g/kg)	88,85	97	113	74	113	83
Silt (g/kg)	59,42	58,42	73	49	77	38
Sand (g/kg)	851,5	844,2	877	817	861	817

Climate

The monthly average historical precipitation and minimum and maximum temperature estimated from a record period of 22 years (1995-2017) (Figure 5), demonstrate that the months of June, July, August, September are the ones that occur the lowest (21.89° C, 21.54° C, 21.16° C,

21.68° C, respectively) and the period from July to November, which presented the lowest precipitation rates (0.51 mm.year-1, 0.33 mm.year^{-1}, 0.16 mm.an-1, 0.67 mm.year^{-1} and 0.78 mm.year^{-1}). Between the months of February and April corresponds to the period that occurs the largest volumes of precipitation, of 8,34 mm.year^{-1}, 10,69 mm.year^{-1} and 10,16 mm.year^{-1}, in this sequence.

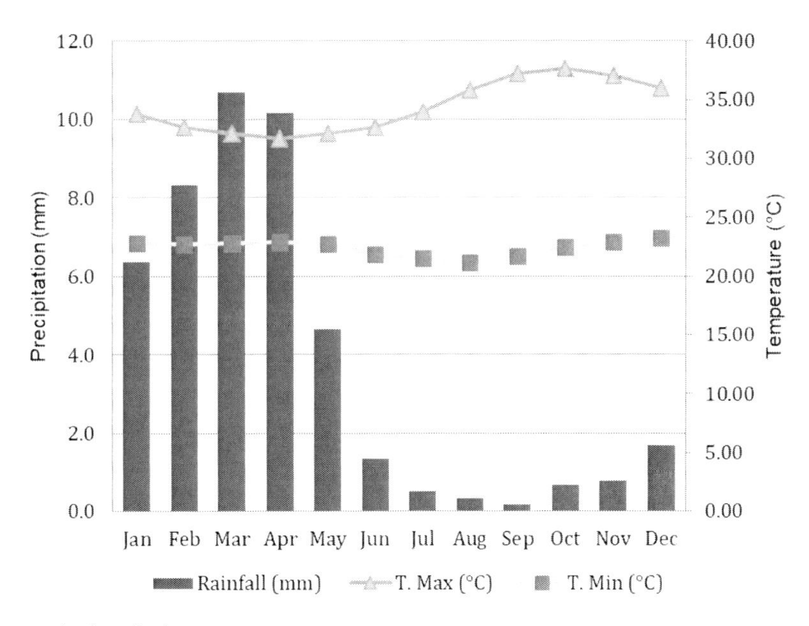

Source: Authors's data.

Figure 5. Climogram of the monthly averages of precipitation and estimated minimum and maximum temperature of a record period of 22 years (1995-2017) in "Palmeira da Emília", Brasileira city, north of Piauí.

Two well-defined periods are characterized in the region, dry and rainy. From February to April corresponds to the rainiest period and the rest of the dry season year. The highest monthly rainfall occurs in March, with 10.69 mm. year^{-1}. The total annual precipitation of the area is around 45.7 mm. year^{-1}, lower than the Carrasco-Caatinga studies: Oliveira et al. (1997); of Carrasco: Araújo et al. (1998); of Caatinga: Alcoforado-Filho (1993), Figuerêdo et al. (2000), Lemos and Rodal (2002), Mendes (2003), Amorim et al. (2005) and Pereira Junior, Andrade and Araújo (2012); of transition Cerrado-Caatinga: Amaral et al. (2012) and Guerra et al. (2014). On the other hand, these values were inferior to two other plant formations of Cerrado: Mesquita and Castro (2007) and Lindoso et al. (2009).

Phytosociology

In the 7,000 m^2 sampled in the phytosociological survey 61 species belonging to 20 botanical families were registered, and 837 individuals were sampled, with a density of 1,195.7 individuals.

The families Fabaceae (nine spp.), Myrtaceae (six spp.), Apocynaceae and Combretaceae (four spp. each) were the most abundant in number of species and, together, accounted for 50.6% of the total individuals (Table 4).

Fabaceae also gained prominence in other surveys of the northeastern semi-arid region, such as Lemos and Rodal (2002) and Lemos and Meguro (2015) in the Caatinga; Lindoso et al. (2010), Mesquita and Castro (2007), Conceição and Castro (2009) and Lima et al. (2013) in the Cerrado; Amaral et al. (2012) in the Cerrado-Caatinga transition area and; Lima, Teodoro and Lemos (2018) in Caatinga-Cerrado transition area.

The Shannon-Wiener diversity index was 2.85 nats $ind.^{-1}$. This value when compared to other surveys in the state of Piauí is inferior to that of Lemos and Rodal (2002) in the Caatinga; Farias and Castro (2004) in transition Cerrado-Caatinga; Mesquita and Castro (2007) and Conceição and Castro (2009) in the Cerrado; Matos and Felfili (2010) in "Matas de Galeria" and Lima, Teodoro and Lemos (2018) in Caatinga-Cerrado transition. On the other hand, they were superior to those of Barbosa et al. (2012) in Catinga in Pernambuco and Amaral et al. (2012) in the Cerrado-Caatinga transition in Piauí.

The lower value of the Shannon-Wiener diversity index may be a consequence of the small number of species with a large percentage of the density, and, according to Alcoforado Filho (1993), this index is very influenced by the density of the dominant species.

Table 4. Families sampled with their phytosociological parameters of a section of vegetation in Brasileira city-Piauí state

Families	Nind	AD	RD	AF	RF	IV	CV
Annonaceae	206	294,3	24,61	100,00	9,72	34,33	24,61
Fabaceae	193	275,7	23,06	100,00	9,72	32,78	23,06
Myrtaceae	128	182,9	15,29	100,00	9,72	25,01	15,29
Indeterminate family	97	138,6	11,59	71,43	6,94	18,53	11,59
Combretaceae	66	94,3	7,89	100,00	9,72	17,61	7,89
Anacardiaceae	21	30,0	2,51	100,00	9,72	12,23	2,51
Apocynaceae	37	52,9	4,42	71,43	6,94	11,36	4,42
Arecaceae	24	34,3	2,87	85,71	8,33	11,20	2,87
Bignoniaceae	23	32,9	2,75	57,14	5,56	8,30	2,75
Vochysiaceae	16	22,9	1,91	57,14	5,56	7,47	1,91
Achariaceae	7	10,0	0,84	28,57	2,78	3,61	0,84
Opiliaceae	4	5,7	0,48	28,57	2,78	3,26	0,48
Rubiaceae	3	4,3	0,36	14,29	1,39	1,75	0,36
Sapotaceae	2	2,9	0,24	14,29	1,39	1,63	0,24
Malpighiaceae	2	2,9	0,24	14,29	1,39	1,63	0,24
Sapindaceae	2	2,9	0,24	14,29	1,39	1,63	0,24
Lamiaceae	2	2,9	0,24	14,29	1,39	1,63	0,24
Ochnaceae	1	1,4	0,12	14,29	1,39	1,51	0,12
Salicaceae	1	1,4	0,12	14,29	1,39	1,51	0,12
Caryocaraceae	1	1,4	0,12	14,29	1,39	1,51	0,12
Dilleniaceae	1	1,4	0,12	14,29	1,39	1,51	0,12
TOTAL	837	1196	100	1028	100	200	100

Nind- number of individuals; AD-absolute density; RD-relative density; AF-absolute frequency; RF-relative frequency; IV - Importance Value; CV-Coverage Value

The maximum, average and minimum heights were 15m, 7.6 m, 3.4 m, respectively (Figure 6). The mean height was higher than that recorded in the surveys of Mesquita and Castro (2007) and Lima et al. (2013), in Cerrado vegetation and; Sousa et al. (2008) in the Cerrado-Caatinga-Carrasco transition area. Height classes with the highest number of individuals were between 5.45-6.45 m, 6.45-7.45 m and 7.45-8.45 m, corresponding to 52.6% of subjects.

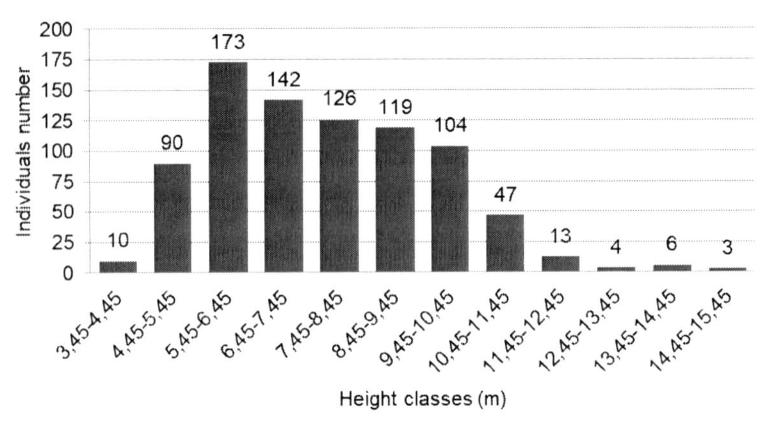

Source: Authors's data.

Figure 6. Distribution of the number of individuals by height classes (m), at fixed intervals of 1 m in a vegetation stretch in "Palmeira da Emília", Brasileira city, North of Piauí.

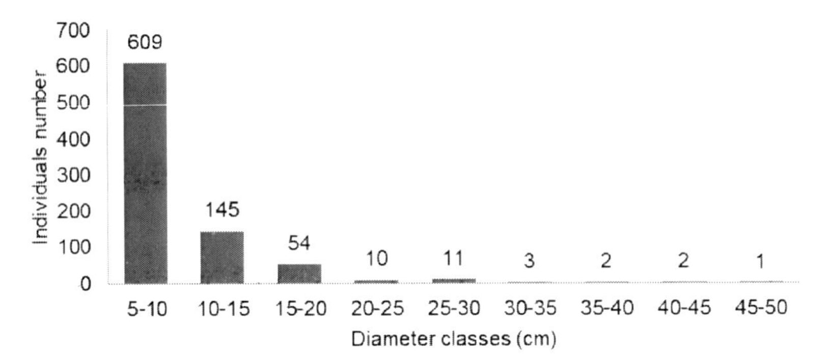

Source: Authors's data.

Figure 7. Distribution of the number of individuals by diameter classes (cm), at fixed intervals of 1 m in "Palmeira da Emília", Brasileira city, north of Piauí.

The maximum, average and minimum diameters were 47,75cm, 9,28cm and 5cm, respectively. This value was higher than Lemos and Rodal (2002) in caatinga vegetation and; smaller than Mendes (2003), also in caatinga vegetation and Farias and Castro (2004), studying an area with several physiognomic types such as "cerrado", "caatinga", "carrasco", "campos" (grasslands) and semi-deciduous forest. In the first diameter classes (Figure 7) there is a high concentration of individuals and this, according to Bertoni (1984), characterizes a vegetation in which the natural regeneration of the species is occurring.

When the number of individuals decreases as the diameter increases, even if there is a disturbance in the vegetation and this community of plants with a smaller diametric size and older will die, young individuals, together with regenerants, will quickly repopulate the affected area (Pereira Júnior, Andrade and Araújo, 2012). In the vegetation sampled, the information of height and mean and maximum diameters suggest a large number of young specimens. This fact probably occurs due to the anthropization of the area. In addition, the large number of sprouts and individuals with small stem diameters at ground level observed in the area may be related to current regeneration processes.

The species identified with higher IV were, decreasingly, *Ephedranthus pisocarpus* R.E.Fr., *Copaifera langsdorffii* Desf., *Myrcia guianensis* (Aubl.) DC., *Terminalia fagifolia* Mart., *Parkia platycephala* (Willd.) Benth. ex Walp., *Astrocaryum campestre* Mart and *Anacardium occidentale* L. (Table 5). These species are also represented with high importance value in the studies of Oliveira (2004), Mesquita and Castro (2007); Sousa et al. (2009) in "cerrado" areas; Santos-Filho (2009) in the area of "restinga" vegetation and Alves et al. (2013) in the "caatinga" area.

Ephedranthus pisocarpus R.E.Fr., *Copaifera langsdorffii* Desf., *Myrcia guianensis* (Aubl.) DC. and *Terminalia fagifolia* Mart., presented the most representative species of IV and CV, with 58.6% of the total area sampled.

Ephedranthus pisocarpus R.E.Fr., according to the Flora of Brazil 2020, may appear in vegetation of "caatinga", riparian forest or "mata de galeria". It was recorded among the species with the highest IV, in the studies carried out by Matos and Felfili (2010) in "matas de galeria" and Sousa et al., (2009) in "cerrado rupestre." *Copaifera langsdorffii* Desf. was the second species with the highest IV and CV in this study. Sarmiento (1983) points to it as one of the common species in the Cerrado and the Flora of Brazil 2020 has its record in the Amazon, Caatinga, Cerrado and Atlantic Forest.

Table 5. Species and their phytosociological parameters of a vegetation stretch in Brasileira city, Piauí state

Species	NInd	AD	RD%	AF	RF%	IV	CV
Ephedranthus pisocarpus R.E.Fr.	205	292,9	24,49	100,00	5,11	29,60	24,49
Copaifera langsdorffii Desf.	146	208,6	17,44	100,00	5,11	22,55	17,44
Myrcia guianensis (Aubl.) DC.	81	115,7	9,68	85,71	4,38	14,06	9,68
Terminalia fagifolia Mart.	59	84,3	7,05	100,00	5,11	12,16	7,05
Parkia platycephala (Willd.) Benth. ex Walp.	18	25,7	2,15	85,71	4,38	6,53	2,15
Astrocaryum campestre Mart	23	32,9	2,75	71,43	3,65	6,40	2,75
Anacardium occidentale L.	14	20,0	1,67	85,71	4,38	6,05	1,67
Indeterminate species 10	32	45,7	3,82	42,86	2,19	6,01	3,82
Myrtaceae	32	45,7	3,82	28,57	1,46	5,28	3,82
Qualea grandiflora Mart.	16	22,9	1,91	57,14	2,92	4,83	1,91
Aspidosperma multiflorum A. DC.	19	27,1	2,27	42,86	2,19	4,46	2,27
Hymenaea courbaril L.	5	7,1	0,60	71,43	3,65	4,25	0,60
Fabaceae	5	7,1	0,60	57,14	2,92	3,52	0,60
Himatanthus drasticus (Mart.) Plumel	4	5,7	0,48	57,14	2.92	3,40	0,48
Handroanthus impetiginosus (Mart. ex DC.) Mattos	10	14,3	1,19	42,86	2,19	3,38	1,19
Indeterminate species 14	8	11,4	0,96	42,86	2,19	3,15	0,96
Indeterminate species 12	14	20,0	1,67	28,57	1,46	3,13	1,67
Indeterminate species 16	20	28,6	2,39	14,29	0,73	3,12	2,39
Handroanthus serratifolius (Vahl) S.Grose	13	18,6	1,55	28,57	1,46	3,01	1,55
Astronium fraxinifolium Schott	5	7,1	0,60	42,86	2,19	2,79	0,60
Combretaceae	4	5,7	0,48	42,86	2,19	2,67	0,48
Fabaceae	10	14,3	1,19	28,57	1,46	2,65	1,19
Campomanesia aromatica (Aubl.) Griseb.	10	14,3	1,19	28,57	1,46	2,65	1,19
Indeterminate species 1	9	12,9	1,08	28,57	1,46	2,54	1,08
Lindackeria ovata (Benth.) Gilg	7	10,0	0,84	28,57	1,46	2,30	0,84
Indeterminate species 20	6	8,6	0,72	28,57	1,46	2,18	0,72
Tachigali vulgaris L. G. Silva & H. C. Lima	5	7,1	0,60	28,57	1,46	2,06	0,60
Indeterminate species 24	4	5,7	0,48	28,57	1,46	1,94	0,48
Indeterminate species 23	4	5,7	0,48	28,57	1,46	1,94	0,48
Agonandra brasiliensis Miers. ex Benth. & Hook. F.	4	5,7	0,48	28,57	1,46	1,94	0,48

Species	NInd	AD	RD%	AF	RF%	IV	CV
Indeterminate species 6	3	4,3	0,36	28,57	1,46	1,82	0,36
Indeterminate species 15	2	2,9	0,24	28,57	1,46	1,70	0,24
Guettarda viburnoides Cham. & Schltdl.	3	4,3	0,36	14,29	0,73	1,09	0,36
Myrtaceae	3	4,3	0,36	14,29	0,73	1,09	0,36
Pouteria ramiflora (Mart.) Radlk.	2	2,9	0,24	14,29	0,73	0,97	0,24
Byrsonima crassifolia (L.) Kunth	2	2,9	0,24	14,29	0,73	0,97	0,24
Magonia pubescens A.St.-Hil.	2	2,9	0,24	14,29	0,73	0,97	0,24
Fabaceae	2	2,9	0,24	14,29	0,73	0,97	0,24
Combretaceae	2	2,9	0,24	14,29	0,73	0,97	0,24
Myracrodruon urundeuva Allemão	2	2,9	0,24	14,29	0,73	0,97	0,24
Vitex flavens Kunth	2	2,9	0,24	14,29	0,73	0,97	0,24
Indeterminate species 2	1	1,4	0,12	14,29	0,73	0,85	0,12
Combretum leprosum Mart	1	1,4	0,12	14,29	0,73	0,85	0,12
Indeterminate species 26	1	1,4	0,12	14,29	0,73	0,85	0,12
Indeterminate species 25	1	1,4	0,12	14,29	0,73	0,85	0,12
Indeterminate species 13	1	1,4	0,12	14,29	0,73	0,85	0,12
Astrocaryum campestre Mart	1	1,4	0,12	14,29	0,73	0,85	0,12
Indeterminate species 21	1	1,4	0,12	14,29	0,73	0,85	0,12
Indeterminate species 19	1	1,4	0,12	14,29	0,73	0,85	0,12
Indeterminate species 18	1	1,4	0,12	14,29	0,73	0,85	0,12
Indeterminate species 17	1	1,4	0,12	14,29	0,73	0,85	0,12
Ouratea hexasperma (A.St.-Hil.) Baill.	1	1,4	0,12	14,29	0,73	0,85	0,12
Indeterminate species 8	1	1,4	0,12	14,29	0,73	0,85	0,12
Indeterminate species 7	1	1,4	0,12	14,29	0,73	0,85	0,12
Casearia grandiflora Cambess.	1	1,4	0,12	14,29	0,73	0,85	0,12
Indeterminate species 3	1	1,4	0,12	14,29	0,73	0,85	0,12
Indeterminate species 9	1	1,4	0,12	14,29	0,73	0,85	0,12
Fabaceae	1	1,4	0,12	14,29	0,73	0,85	0,12
Caryocar coriaceum Wittm	1	1,4	0,12	14,29	0,73	0,85	0,12
Davilla rugosa Poir	1	1,4	0,12	14,29	0,73	0,85	0,12
TOTAL	837	1195	100	1942	100	200	100

NInd-number of individuals; AD-absolute density; RD-relative density; AF- absolute frequency; RF-relative frequency; IV- Importance value; CV-coverage value. The number of undetermined species was given, increasingly, as it appeared in the plots

This species was also recorded in the studies of Alves et al. (2013) in Caatinga and from Lima, Teodoro and Lemos (2018) in Caatinga-Cerrado transition area. *Myrcia guianensis* (Aubl.) DC., a very abundant species in the study area, was also recorded with occurrence in the studies of Sousa et al. (2009), in "cerrado rupestre" and Santos-Filho (2009), in "restinga" vegetation.

Terminalia fagifolia Mart., fourth species with higher IV in this study, was also recorded with higher IV in the studies of Sousa et al. (2009) in "cerrado rupestre" and Matos and Felfili (2010) in "matas de galleria".

In addition to the species that presented higher IV and CV, the importance of the other species that presented low IVs, the presence of few individuals for these species (rare) or the predominance of individuals of small thickness and height, cannot be ignored. These species totaled 29.22% of the species found with only one or two individuals: *Astrocaryum campestre* Mart., *Byrsonima crassifolia* (L.) Kunth. *Caryocar coriaceum* Wittm, *Casearia grandiflora* Cambess., *Combretum leprosum* Mart, *Davilla rugosa* Poir, *Guettarda viburnoides* Cham. & Schltdl., *Magonia pubescens* A.St.-Hil., *Myracrodruon urundeuva* Allemão, *Ouratea hexasperma* (A.St.-Hil.) Baill., *Pouteria ramiflora* (Mart.) Radlk. and *Vitex flavens* Kunth.

The Equability (J ') was 0.696, lower than in some areas of restinga studied by Santos-Filho (2009); of "florestas de galeria" studied by Matos and Felfili (2010) and; in a Caatinga-Cerrado transitional area studied by Lima, Teodoro and Lemos (2018). On the other hand, values close to the study area were recorded in "cerrado rupestre" (Sousa, 2009) and; in the Cerrado-Caatinga-Carrasco transition area (Sousa et al., 2009).

It is worth pointing out this phytosociological study cannot be considered definitive for the total knowledge of the studied plant community, since it was not possible to collect fertile material for a safe scientific identification of some specimens present in the sampling.

FINAL CONSIDERATIONS

With the data obtained from the floristic and phytosociological survey, it is possible to state that the study area also has elements of other plant formations (such as cerrado and caatinga).

Regarding the economic potential of the species, species with different uses (medicinal, food, forage, honey producer, etc.) were observed, predominating species with medicinal and honey producer potential.

Regarding the geographic distribution, it is evident that the northern mesoregion of the state of Piauí has little vegetation catalogued and the study area has a flora that shares characteristics with several plant formations present in the northern Piauiense mesoregion and that this study brings new data for knowledge of local flora and vegetation.

The sampled plant community has high floristic richness and diversity, totaling 837 individuals, comprising 61 species belonging to 20 botanical families. *Ephedranthus pisocarpus* R.E.Fr., *Copaifera langsdorffii* Desf., *Myrcia guianensis* (Aubl.) DC. and *Terminalia fagifolia* Mart., presented the most representative species IV and CV, with 58.6% of the total area sampled.

ACKNOWLEDGMENTS

The authors thank Federal University of Piauí by infrastructure support and by volunteer scholarship (of scientific initiation) to the first author. They also thank Mr. Raimundo Nonato da Silva and Mrs. Benedita Nunes da Silva by hosting at his home during the fieldwork and also the first for the help in the field activities and Mrs. Maria Erismar Nunes da Silva for the logistical support during the development of this research.

References

Agra, M. F. (1996). *Plantas da medicina popular dos Cariris Velhos*, Paraíba, João Pessoa: Ed. União. [*Folk medicine plants of the Old Cariris*]

Albino, R. S. (2005). *Florística e fitossociologia da vegetação de cerrado rupestre de baixa altitude e perfil socioeconômico da atividade mineradora em Castelo do Piauí e Juazeiro do Piauí, Brasil.* Dissertação, Mestrado em Desenvolvimento e Meio Ambiente, Universidade Federal do Piauí, Teresina, PI, Brasil. [*Floristic and phytosociology of cerrado rock vegetation of low altitude and socioeconomic profile of mining activity in Castelo do Piauí and Juazeiro do Piauí, Brazil*]

Alcoforado-Filho, F. G. (1993). *Composição florística e fitossociológica de uma área de caatinga arbórea no Município de Caruaru-PE.* Dissertação de Mestrado Universidade Federal Rural de Pernambuco, Recife, PE, Brasil. [*Floristic and phytosociological composition of an area of arboreal caatinga in the Municipality of Caruaru-PE.]*

Alves, A. R., Ribeiro, I. B., Sousa, J. R. L de, Barros, S. S., Sousa, P da. S. (2013). Análise da estrutura vegetacional em uma área de caatinga no município de Bom Jesus, Piauí. *Revista Caatinga*, 26 (4), 99-106. [Analysis of the vegetation structure in a caatinga area in the municipality of Bom Jesus, Piauí. *Caatinga Magazine*]

Amaral, G. C., Alves, A. R., Oliveira, T. M., Almeida, K. N. S. de, Farias, S. G. G., Botrel, R. T. (2012). Estudo florístico e fitossociológico em uma área de transição Cerrado Caatinga no município de Batalha-PI. *Scientia Plena*, 8, 1-5. [Floristic and phytosociological study in a Cerrado Caatinga transition area in the municipality of Batalha-PI. *Scientia Plena*]

Amaral, M. C. and Lemos, J. R. (2015). Floristic Survey of a Portion of the Vegetation Complex of the Coastal Zone in Piauí State, Brazil. *American Journal of Life Sciences*, 3, 213-218.

Amorim, I. L. de, Sampaio, E. V. S. B., Araújo, E. de L. (2005). Flora e estrutura da vegetação arbustivo-arbórea de uma área de caatinga do

Seridó, RN, Brasil. *Acta Botanica Brasilica*, 19 (3), 615-623. [Flora and structure of the shrub-tree vegetation of a caatinga area of Seridó, RN, Brazil. *Acta Botanica Brasilica*]

Amorozo, M. C. M. (2002). Uso e diversidade de plantas medicinais em Santo Antonio do Leverger, MT, Brasil. *Acta Botanica Brasilica*, 16 (2), 189-203. [Use and diversity of medicinal plants in Santo Antonio do Leverger, MT, Brazil. *Acta Botanica Brasilica*]

Andrade. G. O. (1977). *Alguns aspectos do quadro natural do Nordeste*. Recife: SUDENE. [*Some aspects of the natural picture of the Northeast.*]

Andrade-Lima, D. (1981) The caatingas dominium. *Revista Brasileira de Botânica*, 4, 149-163. [*Brazilian Journal of Botany*]

APG IV. (2016). An update of the Angiosperm Phylogeny Group classification for the orders and families of flowering plants: APG IV. *Botanical Journal of the Linnean Society*, 181, 1-20.

Araújo, F. S., Sampaio, E. V. S. B., Rodal, M. J. N., Figueiredo, M. A. (1998). Organização comunitária do componente lenhoso de três áreas de carrasco em Nova Oriente, CE. *Revista Brasileira de Biologia*, 58 (1), 85-95. [Community organization of the woody component of three areas of hangman in Nova Oriente, CE. *Brazilian Journal of Biology*]

Araujo, J. L., Lemos, J. R. (2015). Estudo etnobotânico sobre plantas medicinais na comunidade de Curral Velho, Luís Correia, Piauí, Brasil. *Revista Biotemas*, 28 (2), 125-136. [Ethnobotanical study on medicinal plants in the community of Curral Velho, Luís Correia, Piauí, Brazil. *Revista Biotemas*]

Baptista, J. G. (1981). *Geografia Física do Piauí*. Teresina: COMEPI. [*Physical Geography of Piauí*]

Baptistel, A. C., Coutinho, J. M. C. P., Lins Neto, E. M. F., Monteiro, J. M. (2014). Plantas medicinais utilizadas na Comunidade Santo Antônio, Currais, Sul do Piauí: um enfoque etnobotânico. *Rev. Bras. Pl. Med*, 16 (2), 406-425. [Medicinal plants used in the Santo Antônio Community, Currais, Southern Piauí: an ethnobotanical approach.]

Barbosa, M. D., Marangon, L. C., Feliciano, A. L. P., Freire, F. J., Duarte, G. M. T. (2012). Florística e fitossociologia de espécies arbóreas e arbustivas em uma área de caatinga em Arcoverde, PE, Brasil. *Revista Árvore*, 36 (5), 851-858. [Floristic and phytosociology of arboreal and shrub species in an area of caatinga in Arcoverde, PE, Brazil. *Home*]

Barbosa, M. R. V. and Peixoto, A. L. (2003). *Coleções botânicas brasileiras: situação atual e perspectivas. Coleções biológicas de apoio ao inventário, uso sustentável e conservação da biodiversidade.* Instituto de Pesquisas Jardim Botânico do Rio de Janeiro, Rio de Janeiro. Pp. 113-125. [*Brazilian botanical collections: current situation and perspectives. Biological collections supporting inventory, sustainable use and biodiversity conservation.* Botanical Garden Research Institute of Rio de Janeiro,]

Barbosa, M. R. V., Mayo, S. J., Castro, A. A. J. F., Freitas, G. L., Pereira, M. S., Gadelha Neto, P. C. and Moreira, H. M. (1996). Checklist preliminar das angiospermas. In: (eds. Sampaio, E. V. S. B., Mayo, S. J. and Barbosa, M. R.) *Pesquisa Botânica Nordestina: Progresso e Perspectivas.* Recife, Sociedade Botânica do Brasil, Seção Regional de Pernambuco, Pp 253-415. [*Northeastern Botanical Research: Progress and Perspectives*]

Barroso, G. M., Guimaraes, E. F. (1980). Excursão botânica ao Parque Nacional de Sete Cidades, Piauí. *Rodriguésia*, 32 (53), 241-267. [Botanical tour to Sete Cidades National Park, Piauí. *Rodrigues*]

Bertoni, J. E. A. (1984). *Composição florística e estrutura de uma floresta do interior do Estado de São Paulo: Reserva Estadual de Porto Ferreira.* 1984. 195p. Dissertação (Mestrado) - Universidade Estadual de Campinas, Unicamp, Campinas. [*Floristic composition and structure of a forest in the interior of the State of São Paulo: State Reserve of Porto Ferreira.*]

Brandon, K., Fonseca, G. A. B., Rylands, A. B., Silva, J. M. (2005). Conservação brasileira: desafios e oportunidades. *Megadiversidade*, 1 (1), 7-13. [Brazilian conservation: challenges and opportunities. *Megadiversity*]

Brazil Flora Group. (2015). Growing knowledge: an overview of Seed Plant diversity in Brazil. *Rodriguésia*, 66 (4), 1085-1113.

Cabral-Freire, M. C. C. and Monteiro, R. (1993). Florística das praias da Ilha de São Luiz, estado do Maranhão (Brasil): diversidade de espécies e sua ocorrência no litoral brasileiro. *Acta Amazonica*, 23, 125-140. [Floristics of the beaches of São Luiz Island, state of Maranhão (Brazil): species diversity and its occurrence in the Brazilian coast.]

Carvalho, E. G. A., Teodoro, M. S., Lemos, J. R. (2018). Inventario florístico de uma área ecotonal caatinga-cerrado no Norte do Piauí, Nordeste do Brasil. In: Lemos, J. R. (Org). *Pesquisas Botânicas e Ecológicas no Piauí*. Curitiba: CRV, 2018. Coedição: Teresina, PI: EDUFPI, Pp. 35-54. [Floristic inventory of a caatinga-cerrado ecotonal area in Northern Piauí, Northeast Brazil. In: *Botany and Ecological Research in Piauí*.]

Castro, A. A. J. F. (1994). *Composição florístico-geográfica (Brasil) e fitossociológica (Piauí-São Paulo) de amostras de cerrado*. Tese de Doutorado em Ciências, Universidade Estadual de Campinas, Campinas. [*Floristic-geographic composition (Brazil) and phytosociological (Piauí-São Paulo) of cerrado samples*.]

Castro, A. A. J. F., Castro, A. S. F., Farias, R. R. S., Sousa, S. R. de, Castro, N. M. C. F., Silva, C. G. B., Mendes, M. R. A., Barros, J. S., Lopes, R. N. (2009). Diversidade de espécies e de ecossistemas da vegetação remanescente da Serra Vermelha, área de chapada, municípios de Curimatá, Redenção do Gurguéia e Morro Cabeça no Tempo, Sudeste do Piauí. *Publicações Avulsas Conservação do Ecossistema*, 23, 1-72. [Diversity of species and ecosystems of the remnant vegetation of the Serra Vermelha, plateau area, Curimatá, Redenção do Gurguéia and Morro Cabeça no Tempo, Southeast of Piauí. *Conservation of the Ecosystem*]

Castro, A. A. J. F., Farias, R. R. S., Sousa, S. R., Castro, N. M. C. F., Barros, J. S., Lopes, R. N. (2014). Caracterização florísticas e estrutura da comunidade arbórea de um remanescente de floresta estacional, município de Manoel Emídio e Alvorada da Gurguéia, Piauí, Brasil. *Publicação Avulsas Conservação de Ecossistemas*, Teresina, 23, 1-18. [Floristic characterization and structure of the tree community of a remnant of seasonal forest, municipality of Manoel Emídio and

Alvorada da Gurguéia, Piauí, Brazil. Publication Avulsas *Conservation of Ecosystems*]

Castro, A. A. J. F., Martins, F. M., Fernandes, A. G. (1998). The woody flora of cerrado vegetation in the state of Piauí, northeastern Brazil. *Edinb. J. Bot.*, 55 (3), 455-472.

CEPRO (1996). Fundação Centro de Pesquisas Econômicas e Sociais do Piauí. *Piauí: caracterização do quadro natural.* CEPRO: Teresina. [Foundation Center for Economic and Social Research of Piauí. *Piauí: characterization of the natural picture*]

Cerqueira, E. C. and Lemos, J. R. (2018). Levantamento florístico em trilhas naturais de um sitio com potencial turístico no norte do Piauí como subsídio à educação ambiental e conservação da fitodiversidade. In: Lemos, J. R. (Org). *Pesquisas Botânicas e Ecológicas no Piauí.* Curitiba: CRV, Coedição: Teresina, PI: EDUFPI, Pp. 149-159. [Floristic survey on natural trails of a site with tourist potential in the north of Piauí as a subsidy to environmental education and conservation of phytodiversity. In: *Botany and Ecological Research in Piauí.*]

Chaves, E. M. F. (2005). *Florística e potencialidades econômicas da vegetação de carrasco no município de Cocal, Piauí, Brasil.* Dissertação de Mestrado em Desenvolvimento e Meio Ambiente, Universidade Federal do Piauí, Teresina-PI, Brasil. [*Floristics and economic potential of hangman vegetation in the municipality of Cocal, Piauí, Brazil*]

Chaves, E. M. F., Barros, R. F. M. de, Araújo, F. S. de. (2007). Flora Apícola do Carrasco no Município de Cocal, Piauí, Brasil. *Revista Brasileira de Biociências*, Porto Alegre, 5, 555-557. [Carrasco Apiculture Flora in the Municipality of Cocal, Piauí, Brazil. *Brazilian Journal of Biosciences*]

Conceição, G. M. da., Castro, A. A. J. F. (2009). Fitossociologia de uma área de cerrado marginal, Parque Estadual do Mirador, Mirador, Maranhão. *Scientia Plena*, 5 (10), 1-16. [Phytosociology of an area of marginal cerrado, Mirador State Park, Mirador, Maranhão. *Scientia Plena*]

Costa, J. A. S., Nunes, T. S., Ferreira, A. P. L. Stradmann, M. T. S., Lorenzi, H., Matos, F. J. A. (2002). *Plantas Medicinais no Brasil – nativas e exóticas.* Nova Odessa (SP): Instituto Plantarum, Pp 512. [*Medicinal Plants in Brazil - native and exotic*]

Costa, J. L. P de. O., Cavalcanti, A. P. B. (2010). Fitogeografia da planície deltaica do rio Parnaíba, Piauí/Maranhão-Brasil: Análise da distribuição das espécies e interferência antrópica. *Observatorium: Revista Eletrônica de Geografia*, 2 (4), 84-92. [Phytogeography of the Parnaíba River Delta Plain, Piauí / Maranhão-Brazil: Analysis of species distribution and anthropic interference. *Observatorium: Electronic Journal of Geography*]

Costa, J. M. (2005). *Estudo fitossociológico e socioambiental de uma área de Cerrado com potencial melitófilo no município de Castelo do Piauí, Piauí, Brasil.* Dissertação de Mestrado em Desenvolvimento e Meio Ambiente, Universidade Federal do Piauí, Teresina-PI, Brasil. [*Phytosociological and socioenvironmental study of a Cerrado area with potential melitophilus in the municipality of Castelo do Piauí, Piauí, Brazil.*]

Eiten, G. (1977). *Delimitação do conceito de cerrado*. Arquivos do Jardim Botânico do Rio de Janeiro, 21, 125-134. [*Delimitation of the concept of closed*]

Engel, V. L., Fonseca, R. C. B., Oliveira, R. E de. (1998). Ecologia de lianas e o manejo de fragmentos florestais. *Série Técnica IPEF*, 12 (32), 43-64. [Ecology of lianas and the management of forest fragments. *IPEF Technical Series*]

Faleiro F. G., Gama L, C., Farias Neto A. L., Sousa E. S., (2008). *O* Simpósio Nacional sobre o Cerrado e o Simpósio Internacional sobre Savanas Tropicais. In: Faleiro, F. G. and Farias Neto, A. L. de (eds.). *Savanas: desafios e estratégias para o equilíbrio entre sociedade, agronegócio e recursos naturais.* Planaltina: Embrapa Cerrados, Pp. 33-48. [National Symposium on the Cerrado and the International Symposium on Tropical Savannas. In: *Savanas: challenges and strategies for the balance between society, agribusiness and natural resources.*]

Farias, R. R. S. (2003). *Florística e fitossociologia em trechos de vegetação do complexo de Campo Maior, Campo Maior, Piauí.* Dissertação de Mestrado em Biologia Vegetal, Universidade Federal de Pernambuco, Recife, Brasil. [*Floristic and phytosociology in vegetation stretches of the Campo Maior complex, Campo Maior, Piauí.*]

Farias, R. R. S., Castro, A. A. J. F. (2004). Fitossociologia de trechos da vegetação do Complexo Campo Maior, PI, Brasil. *Acta botanica brasilica*, 18 (4), 949-963. [Phytosociology of vegetation stretches of Campo Maior Complex, PI, Brazil.]

Fernandes, A. (2003). *Conexões florísticas do Brasil.* Fortaleza: Banco do Nordeste, Pp 135. [*Floral connections of Brazil*]

Fernandes, A. G. (1982). A vegetação do Piauí. In: *Congresso Nacional De Botânica*, 32. 1981, Teresina. Anais... Teresina: Sociedade Botânica do Brasil, Pp 313-318. [The vegetation of Piauí. In: *National Congress of Botany*]

Ferraz, E. M. N., Araújo, E de. L., Silva, K. A da. (2009). Estudo florístico do componente herbáceo e relação com solos em áreas de caatinga do embasamento cristalino e bacia sedimentar, Petrolândia, PE, Brasil. *Acta bot. bras.*, 23 (1), 100-110. 2009. [Floristic study of the herbaceous component and relation with soils in caatinga areas of the crystalline basement and sedimentary basin, Petrolândia]

Figueirêdo, L. S., Rodal, M. J. N., Melo, A. L. (2000). Florística e fitossociologia de uma área de vegetação arbustiva caducufólia no município de Buíque-Pernambuco. *Naturalia*, 25, 205-224. [Floristic and phytosociology of an area of shrubby vegetation in the municipality of Buíque-Pernambuco.]

Flora do Brasil 2020 em Construção. (2018). *Jardim Botânico do Rio de Janeiro.* Acesso em 4 Ago, Disponível em: http://floradobrasil.jbrj.gov.br. [*Botanical Garden of Rio de Janeiro.*]

Fonseca, M. R. (1991). *Análise da vegetação arbustivo-arbóreo da Caatinga hiperxerófila do Nordeste do Estado de Sergipe*. Tese de Doutorado em Ciências, Universidade Estadual de Campinas, SP,

Brasil. [*Analysis of the shrub-arboreal vegetation of the hyperoxerophilic Caatinga of Northeast of the State of Sergipe*.]

Forzza, R. C., Baumgratz, J. F. A., Bicudo, C. E. M., Canhos, D. A. L., Carvalho Jr, A. A., Coelho, M. A. N., Costa, A. F., Costa, D. P., Hopkins, M. G., Leitman, P. M., Lohmann, L. G., Lughadha, E. N., Maia, L. C., Martinelli,G., Menezes, M., Morim, M. P., Peixoto, A. L., Pirani, J. R., Prado, J., Queiroz, L. P., Souza, S., Souza, V. C., Stehmann, J. R., Sylvestre, L. S., Walter, B. M. T., Zappi, D. C. (2012). New Brazilian Floristic List Highlights Conservation Challenges. *BioScience*, 62 (1), 39-45.

Foury, A. P. (1972). As matas do Nordeste brasileiro e sua importância econômica. *Boletim de Geografia*, 31 (227), 14-131. [The forests of the Brazilian Northeast and its economic importance. *Bulletin of Geography*]

Franco, E. A. P., Barros, R. F. M. (2006). Uso e diversidade de plantas medicinais no Quilombo Olho D'água dos Pires, Esperantina, Piauí. *Revista Brasileira de Plantas Medicinais*, 8 (3), 78-88. [Use and diversity of medicinal plants in Quilombo Olho D'água dos Pires, Esperantina, Piauí. *Brazilian Journal of Medicinal Plants*]

Giulietti, A. M., Rapini, A., Andrade, M. J. G., Queiroz, L. P., Silva, J. M. C. (2009). *Plantas raras do Brasil*. Belo Horizonte, Conservação Internacional. [*Rare plants of Brazil*.]

Gomes, M. A. F. (1979). *Padrões de Caatinga nos Cariris Velhos, Paraíba*. Dissertação de Mestrado em Botânica, Universidade Federal Rural de Pernambuco, Recife, PE, Brasil. [*Patterns of Caatinga in Cariris Velhos, Paraíba.*]

Guerra, A. M. N. de M., Pessoa, M. de F., Maracajá, P. B. (2014). Estudo fitossociológico em dois ambientes da caatinga localizada no assentamento Moacir Lucena, Apodi-RN – Brasil. *Revista Verde de Agroecologia e Desenvolvimento Sustentável*, 9 (1), 141 -150. [Phytosociological study in two environments of the caatinga located in the settlement Moacir Lucena, Apodi-RN - Brazil. *Green Magazine on Agroecology and Sustainable Development*]

Haidar, R. F. (2008). *Fitossociologia, diversidade e sua relação com variáveis ambientais em florestas estacionais do bioma Cerrado no planalto central e Nordeste do Brasil.* Dissertação de Mestrado em Ciências Florestais, Universidade de Brasília, Brasília, Brasil. [*Phytosociology, diversity and its relationship with environmental variables in seasonal forests of the Cerrado biome in the central and Northeastern plateau of Brazil*]

Haidar, R. F., Felfili, J. M., Matos, M. Q., Castro, A. A. J. F. (2010). Fitossociologia e diversidade da comunidade arbórea de floresta estacional semidecidual do Parque Nacional de Sete Cidades (Piauí) e sua correlação florística com outras florestas estacionais do Brasil. In: Castro, A. A. J. F., Arzabe, C., Castro, N. M. C. F. (Orgs.). *Biodiversidade e ecótonos da região setentrional do Piauí.* Teresina: EDUFPI, Desenvolvimento e Meio Ambiente 5, 141-165. [Phytosociology and diversity of the semideciduous seasonal forest tree community of the Sete Cidades National Park (Piauí) and its floristic correlation with other seasonal forests of Brazil. In: *Biodiversity and ecotones of the northern region of Piauí.*]

Harley, R. M., Atkins, S., Budantsev, A., Cantino, P. H., Conn, B., Grayer, R., Harley, M. M., Kok, R., Krestovskaja, T., Morales, A., Paton, A. J., Ryding, O., Upson, T. (2004). Labiatae. In: Kadereit, J. W. (ed.). *The families and genera of vascular plants*, Kubitzki, K., ed., 7, 167-275.

Instituto Brasileiro de Geografia e Estatística. *Censo Demográfico 2010. (2017). Retratos do Brasil e do Piauí, 2011.* Acesso 20 set, Disponível em: http://www.ibge.gov.br/home/presidencia/noticias/pdf/censo_2010_piaui.pdf. [*Demographic Census 2010. (2017). Portraits of Brazil and Piauí, 2011.*]

Instituto Brasileiro de Geografia e Estatística. 2004. (2017). Acesso em: 21 de dezembro de 2017, [Accessed on: December 21, 2017] Disponível em: http://www.ibge.gov.br.

Klink, C. A., Machado, R. B. (2005). A conservação do cerrado brasileiro. *Megadiversidade.* 1 (1), 147-155. [The conservation of Brazilian cerrado. *Megadiversity*]

Lawrence, G. H. M. (1973). *Taxonomia das plantas vasculares* (vol. 2, Pp. 256). Lisboa: Fundação Calouste Gulbenkian. [*Taxonomy of vascular plants*]

Lemos, J. R. (2004). Composição florística do Parque Nacional Serra da Capivara, Piauí, Brasil. *Rodriguésia*, 55, 55-66. [Floristic composition of Serra da Capivara National Park, Piauí, Brazil. *Rodrigues*]

Lemos, J. R., Meguro, M. (2015). Estudo fitossociológico de uma área de Caatinga na Estação Ecológica (ESEC) de Aiuaba, Ceará, Brasil. *Biotemas*, 28 (2), 39-50. [Phytosociological study of a Caatinga area at the Ecological Station (ESEC) of Aiuaba, Ceará, Brazil. *Biotemas*]

Lemos, J. R., Rodal, M. J. N. (2002). Fitossociologia do componente lenhoso de um trecho de vegetação de Caatinga no Parque Nacional Serra da Capivara, Piauí, Brasil. *Acta Botanica Brasilica*, 16 (1), 23-42. [Phytosociology of the woody component of a vegetation stretch of Caatinga in Serra da Capivara National Park, Piauí, Brazil. *Acta Botanica Brasilica*]

Lewington, A. (1990). *Plants for people*. London: Natural History Museum Publ.

Lewinsohn, T. M., Prado, P. I. (2005). Quantas espécies há no Brasil? *Megadiversidade*. 1 (1), 36-42. [How many species are there in Brazil? *Megadiversity*.]

Lima, E. G. N., Alencar, N. L., Lopes, C. G. R. (2013). Levantamento fitossociológico de uma área de cerradão da fazenda experimental do Colégio Agrícola, Floriano (PI). In: *Congresso Nacional de Botânica 64. Belo Horizonte, Resumos... Belo Horizonte*. [Phytosociological survey of a cerradão area of the experimental farm of the Agricultural College, Floriano (PI). In: *National Congress of Botany 64. Belo Horizonte, Abstracts ... Belo Horizonte*]

Lima, G. A., Teodoro, M. S, Lemos, J. R. (2018). Estrutura de um trecho de vegetação subcaducifólia no extremo Norte do Piauí, Brasil. In: Lemos, J. R. (Org). *Pesquisas Botânicas e Ecológicas no Piauí.* Curitiba, CRV, Coedição: Teresina, PI: EDUFPI. [Structure of a stretch of subcaducifolia vegetation in the extreme north of Piauí, Brazil. In: *Botany and Ecological Research in Piauí.*]

Lima, M. M., Monteiro, R., Castro, A. A. J. F., Costa, J. M. (2010). Levantamento florístico e fitossociológico do morro do Cascudo, área de entorno do Parque Nacional de Sete Cidades (PN7C), Piauí, Brasil. In: Castro, A. A. J. F., Arzabe, C., Castro, N. M. C. F. (Orgs.). *Biodiversidade e ecótonos da região setentrional do Piauí.* Teresina: EDUFPI. Pp. 186-207. [Floristic and phytosociological survey of Cascudo hill, area surrounding the Sete Cidades National Park (PN7C), Piauí, Brazil. In: *Biodiversity and ecotones of the northern region of Piauí.*]

Lindoso, G. da S., Felfili, J. M., Costa, J. M da, Castro, A. A. J. F. (2009). Diversidade e estrutura do cerrado *sensu stricto* sobre areia (Neossolo Quartzarênico) na Chapada Grande Meridional, Piauí. *Rev. Biol. Neotrop*, 6 (2), 45-61. [Diversity and structure of cerrado sensu stricto on sand (Quartzarênico Neosol) in Chapada Grande Meridional, Piauí.]

Lindoso, G. S., Felfini, J. M., Castro, A. A. J. (2010). Diversidade e estrutura do Cerrado *sensu stricto* sobre Areias (Neossolo Quartzarênio) no Parque Nacional de Sete Cidades (PN7C), Piauí. In: Castro, A. A. J. F., Arzabe, C., Castro, N. M. C. F. (Orgs.). *Biodiversidade e ecótonos da região setentrional do Piauí.* Teresina: EDUFPI, Desenvolvimento e Meio Ambiente 5, Pp 186-207. [Diversity and structure of Cerrado sensu stricto on Sands (Neosol Quartzarênio) in the National Park of Sete Cidades (PN7C), Piauí. In: *Biodiversity and ecotones of the northern region of Piauí.*]

Lorenzi, H., Souza, H. M., Torres, M. A. V., Bacher, L. B. (2003). *Árvores exóticas no Brasil – madeireiras, ornamentais e aromáticas.* Nova Odessa (SP): Instituto Plantarum, Pp. 368. [*Exotic trees in Brazil - logging, ornamental and aromatic.*]

Maia, G. N. (2004). *Caatinga – árvores e arbustos e suas utilidades.* 1. ed. São Paulo: D & Z, Pp. 413. [*Caatinga - trees and shrubs and their utilities.*]

Martinelli, G., Moraes, M. A. (2013). *Livro vermelho da flora do Brasil.* [*Red flora book of Brazil.*] Instituto de Pesquisas Jardim Botânico do Rio de Janeiro, Pp. 1100.

Matos, M. Q. (2009). *Matas de galeria no Parque Nacional de Sete Cidades (PNSC), Piauí, Brasil: fitossociologia, diversidade, regeneração natural e relação com variáveis ambientais*, Dissertação de Mestrado em Ciências Florestais, Universidade de Brasília, Brasília. [*Gallery woods in the Sete Cidades National Park (PNSC), Piauí, Brazil: phytosociology, diversity, natural regeneration and relation with environmental variables*]

Matos, M. Q., Felfili, J. M. (2010). Florística, fitossociologia e diversidade da vegetação arbórea nas matas de galeria do Parque Nacional de Sete Cidades (PNSC), Piauí, Brasil. *Acta Botanica Brasilica*, 24 (2), Pp. 483-496. [Floristic, phytosociology and diversity of tree vegetation in gallery forests of Sete Cidades National Park (PNSC), Piauí, Brazil.]

Mendes, M. R. A. (2003). *Florística e fitossociologia de um fragmento de caatinga arbórea, São José do Piauí, Piauí.* Dissertação de Mestrado em Biologia Vegetal, Universidade Federal de Pernambuco, Recife, PE, Brasil. [*Floristic and phytosociology of a fragment of arboreal caatinga, São José do Piauí, Piauí.*]

Mendes, M. R. A., Castro, A. A. J. F. (2010). Vascular Flora of Semi-Arid Region, São José de Piauí, Stat of Piauí Brazil. *Check List*, 6 (1), Pp. 39-44.

Mendes, M. R. A., Munhoz, C. B. R., Júnior, M. C. S., Castro, A. A. J. F. (2012). Relação entre a vegetação e as propriedades do solo em áreas de campo limpo úmido no Parque Nacional de Sete Cidades, Piauí, Brasil. *Rodriguésia*, 63 (4), Pp. 971-984. [Relation between vegetation and soil properties in clean wetland areas in the Sete Cidades National Park, Piauí, Brazil. *Rodrigues*]

Mesquita, M. R., Castro, A. A. J. F. (2007). Florística e fitossociologia de uma área de Cerrado marginal (Cerrado baixo), Parque Nacional Sete Cidades, Piauí. *Publicação Avulsas Conservação de Ecossistemas*, 15, Pp. 1-22. [Floristic and phytosociology of a marginal Cerrado area (Cerrado low), Sete Cidades National Park, Piauí. *Conservation of Ecosystems*]

MMA. Ministério do Meio Ambiente. *O Bioma Cerrado*, Acesso em: 7 de Fevereiro de 2018. Disponível em: http://www.mma.gov.br. [*The Closed Biome*]

Mori, S. A., Silva, L. A. M., Lisboa, G., Coradin, L. (1989). *Manual de manejo do herbário fanerogâmico.* 2 ed. Ilhéus, Bahia: Centro de Pesquisas do Cacau, Pp. 103. [*Handbook of the phanerogamic herbarium*]

Moura, I. O., Felfini, J. M., Pinto, J. R. R., Castro, A. A. J. (2010). Composição florística e estrutura do componente lenhoso em cerrado *sensu stricto* sobre afloramento rochosos no Parque Nacional de Sete Cidades- PI. In: Castro, A. A. J. F., Arzabe, C., Castro, N. M. C. F. (Orgs.). *Biodiversidade e ecótonos da região setentrional do Piauí.* Teresina: EDUFPI, (Desenvolvimento e Meio Ambiente 5, Pp. 90-115. [Floristic composition and structure of the woody component in cerrado sensu stricto on rocky outcrop in the Sete Cidades-PI National Park. In: *Biodiversity and ecotones of the northern region of Piauí.*]

Myers, N., Mittermeier, C. G., Fonseca, G. A. B., Kent, J. (2000). Biodiversity hotspots for conservation priorities. *Nature*, 403, 853-858.

Oliveira, F. C. S, Barros, R. F. M, Moita Neto, J. M. (2010). Plantas medicinais utilizadas em comunidades rurais do semiárido piauiense. *Revista Brasileira de Plantas Medicinais*, 12 (3), Pp. 282-301. [Medicinal plants used in rural communities in the semi-arid region of Piauí. *Brazilian Journal of Medicinal Plants*]

Oliveira, L. S. D., Soares, S. M. N. A., Soares, F. A R., Barros, R. F. M. (2007). Levantamento Florístico do Parque Ambiental Paquetá, Batalha, Piauí. *Revista Brasileira de Biociências*, Porto Alegre, 5, 372-374. [Floristic survey of the Paquetá Environmental Park, Batalha, Piauí. *Brazilian Journal of Biosciences*]

Oliveira, M. E. A. (2004). *Mapeamento, florística e estrutura da transição campo floresta na vegetação (Cerrado) do Parque Nacional de Sete Cidades, Nordeste do Brasil.* Tese de Doutorado, Universidade Estadual de Campinas, São Paulo, SP, Brasil. [*Mapping, floristic and transition structure forest field in the vegetation (Cerrado) of the National Park of Sete Cidades, Northeast of Brazil.*]

Oliveira, M. E. A., Sampaio, E. V. B., Castro, A. A. J. Rodal, M. J. N. (1997). Flora e fitossociologia de uma área de transição Carrasco-caatinga de areia em Padre Marcos, Piauí. *Naturalia*, 22, 131-150. [Flora and phytosociology of a transition area Carrasco-caatinga sand in Padre Marcos, Piauí.]

Pereira Júnior, L. R., Andrade, A. P., Araújo, K. D. (2012). Composição florística e fitossociológica de um fragmento de caatinga em Monteiro, PB. *Holos*, 6, 72-87. [Floristic and phytosociological composition of a caatinga fragment in Monteiro, PB.]

Pereira, V. S., Lemos, J. R. (2018). Levanameno florístico no povoado Pontal do Anel, Luís Correia, Piauí, Nordeste do Brasil. In: Lemos, J. R. (Org). *Pesquisas Botânicas e Ecológicas no Piauí.* Curitiba: CRV, 2018. Coedição: Teresina, PI: EDUFPI, p. 123-147. [Floristic levanameno in the town of Pontal do Anel, Luís Correia, Piauí, Northeast of Brazil. In: *Botany and Ecological Research in Piauí.*]

Pessoa, L. M., Santos-Filho, F. S. (2011). Florística e estrutura do estrato herbáceo em cinco municípios do Estado do Piauí. In: Santos-Filho, F. S., Soares, A. F. C. L. (Org). *Biodiversidade do Piauí.* 1. Ed. Curitiba, PR: CVR, Pp. 199. [Floristics and structure of the herbaceous stratum in five municipalities of the State of Piauí. In: *Biodiversity of Piauí*]

Rapini, A. (2004). *Sistemática: Estudos em Asclepiadoideae (Apocynaceae) da Cadeia do Espinhaço-Minas Gerais.* Tese de Doutorado, Instituto de Biociências da Universidade de São Paulo, São Paulo, SP, Brasil. [*Systematics: Studies in Asclepiadoideae (Apocynaceae) of the Chain of Espinhaço-Minas Gerais.*]

Ribeiro, J. F., Sano, S. M. E. da Silva, J. A. (1981). Chave preliminar de identificação dos tipos fisionômicos da vegetação do Cerrado. In: *Anais do XXXII Congresso Nacional de Botânica*. Sociedade Botânica do Brasil, Teresina, Brasil, Pp. 124-133. [Preliminary key of identification of the physiognomic types of the Cerrado vegetation. In: *Proceedings of the 32nd National Congress of Botany*. Brazilian Society of Botany]

Ribeiro, J. F. and Walter, B. M. T. (1998). Fitofisionomias do Bioma Cerrado. In: Sano, S. M., Almeida, S. P. (Eds.). *Cerrado: ambiente e flora*. Planaltina, EMBRAPA, Pp. 89-166. [Phytophysiognomies of the Cerrado Biome. In: Sano, S.M., Almeida, S.P. (Eds.). *Closed: environment and flora*]

Rizzini, C. I. and Mors, W. G. (1995). *Botânica econômica brasileira.* 2. ed. São Paulo: EPU – EDUSP, Pp. 242. [*Brazilian economic botany*]

Rocha, A. M., Luz, A. R. M., Abreu, M. C. de A. (2017). Composição e similaridade florística de espécies arbóreas em uma área de caatinga, Picos, Piauí. *Pesquisas/Botânica*, Rio Grande do Sul, 70, Pp. 175-185. [Composition and floristic similarity of tree species in an area of caatinga, Picos, Piauí. *Research / Botany*]

Rodal, M. J. N. (1992). *Fitossociologia da vegetação arbustivo-arbórea em quatro áreas de caatinga em Pernambuco.* Tese de Doutorado em Ciências. Universidade Estadual de Campinas, São Paulo, SP, Brasil. [*Phytosociology of shrub-tree vegetation in four caatinga areas in Pernambuco*]

Rodrigues, S. M. C. B. (1998). *Florística e fitossociologia de uma área de cerrado em processo de desertificação no município de Gilbués-PI.* Dissertação de Mestrado, Universidade Federal Rural de Pernambuco, Recife, PE, Brasil. [*Floristic and phytosociology of an area of cerrado in the process of desertification in the municipality of Gilbués-PI.*]

Sampaio, E. V. S. B. (1996). Fitossociologia. In: Sampaio, E. V. S. B, Mayo, S. J., Barbosa, M. R. V. (Eds.) *Pesquisa botânica nordestina: progresso e perspectivas.* Recife: Sociedade Botânica do Brasil. Pp. 203-224. [*Northeastern Botanical Research: Progress and Perspectives*]

Santos, D. A., Andrade, I. M., Araujo, J., Lemos, J. R. (2017). Chave de identificação de caracteres vegetativos do estrato arbóreo-arbustivo de um trecho na Zona Urbana no Norte do Piauí. *Espacios (Caracas)*, 38, 7-14. [Identification key of vegetative characters of the arboreal-shrub stratum of a stretch in the Urban Zone in the North of Piauí. *Spaces (Caracas)*]

Santos, M. F. A. V. (1987). *Características de solo e vegetação em sete áreas de Parnamirim, Pernambuco*. Dissertação de Mestrado em Botânica, Universidade Federal Rural de Pernambuco, Recife, PE,

Brasil. [*Soil and vegetation characteristics in seven areas of Parnamirim, Pernambuco*]

Santos-Filho, F. S., Almeida Jr, E. B. de., Lima, P. B., Soares, C. J. dos R. A. (2015). Checklist of the flora of the restingas of Piauí state, Northeast Brazil. *Check List* 11(2), 1-10.

Santos-Filho, F. S. (2009). *Composição florística e estrutural da vegetação de restinga do Estado do Piauí*, Tese de Doutorado em Botânica, Universidade Federal Rural de Pernambuco, Departamento de Biologia, Recife, PE, Brasil. [*Floristic and structural composition of the restinga vegetation of the State of Piauí.*]

Santos-Filho, F. S., Mesquita, T. K. da S., Almeida Jr, E. B., Zickel, C. S. de. (2016). A flora de Cajueiro da Praia: uma área de Tabuleiros do Litoral do Piauí, Brasil. *Revista Equador* (UFPI), 5 (2), Pp. 21-35. [The flora of Cajueiro da Praia: an area of Tabuleiros do Litoral do Piauí, Brazil.]

Sarmiento, G. (1983). The savannas of tropical America. In *Tropical savannas. Ecossystems of the world.* 13 (F. Bouliére, ed.). Elsevier Science Publishers, pp. 245-288.

Shafer, C. L. (1990). *Nature reserves: island theory and conservation practice.* Washington: Smithsonian Institution Press, Pp. 189.

Silva, A. K. C. da, Lemos, J. R. (2018). Florística de uma área de transição no Norte do Piauí, Nordeste do Brasil. In: Lemos, J. R. (Org). *Pesquisas Botânicas e Ecológicas no Piauí*. Curitiba: CRV, Coedição: Teresina, PI: EDUFPI, Pp. 13-33. [Floristics of a transition area in the North of Piauí, Northeast of Brazil. In: *Botany and Ecological Research in Piauí*.]

Silva, A. M de. S., Silva, D. F. M da., Sousa, G. M de. (2017). Composição Florística no Nazareth Eco Resort, município de José de Freitas – PI. *Educação ambiental em ação*. N. 61, Ano XVI. [Floristic Composition at Nazareth Eco Resort, municipality of José de Freitas - PI. *Environmental education in action.*]

Silva, A. T., Muniz, C. F. S., Wanderley, M. G. L., Kirizawa, M., Sendulsky, T., Silva, T. S., Maluf, A. M., Silvestre, M. S. F., Chiea, S. A. C., Custódio-Filho, A., Mantovani, W., Jung, S. L., Barros, F.,

Oliveira, L. C. A. (1989). Pteridófitas e fanerógamas. In: Fidalgo, O., Bononi, V. L. R. *Técnicas de coleta, preservação e herborização de material botânico.* Série Documentos. São Paulo: Instituto de Botânica, Pp. 62. [Pteridophytes and phanerogams. In: *Collection, preservation and herborization techniques of botanical material.*]

Silva, G. A. dos R., Bastos, E. M., Sobreira, J. A. dos R. (2014). Levantamento da flora apícola em duas áreas produtoras de mel no Estado do Piauí. *Enciclopédia Biosfera,* Centro Científico Conhecer - Goiânia, 10 (18), Pp. 3305-3316. [Survey of the bee flora in two honey producing areas in the State of Piauí. *Encyclopedia Biosphere.*]

Silva, J. M. C., Bates, J. M. (2002). Biogeographic Patterns and Conservation in the South American Cerrado: A Tropical Savanna Hotspot. *BioScience,* 52 (3), 225-234.

Silva, L. S da., Alves A. R., Nunes, A. K. A., Macedo, W. de S., Martins A. da R. (2015). Florística, estrutura e sucessão ecológica de um remanescente de mata ciliar na bacia do rio Gurguéia-PI. *Nativa,* 3 (3), 156-164. [Floristic, structure and ecological succession of a remnant of riparian forest in the Gurguéia-PI basin.]

Silva, M. P. da. (2010). *Etnobotânica de comunidades rurais da serra de Campo Maior- PI, Brasil.* Dissertação de Mestrado em Desenvolvimento e Meio Ambiente, Universidade Federal do Piauí, Teresina-PI, Brasil. [*Ethnobotany of rural communities in the mountain range of Campo Maior- PI, Brazil.*]

Simpson, B. B. and Ogorzaly, M. C. (1995). *Plants in our world.* 2nd. New York: Ed. McGraw-Hill.

Sobral, L. F., Barretto, M. C de V., Silva, A. J. da S., Anjos, J. L. dos. (2015). *Documentos 206-Guia Prático para Interpretação de Resultados de Análises de Solo.* Aracaju: Embrapa Tabuleiros Costeiros, Pp. 13. [*Documents 206-Practical Guide to Interpretation of Soil Analysis Results.*]

Sousa, D. M. G. de., Lobato, E. (2004). *Cerrado: correção do solo e adubação.* Embrapa Informação Tecnológica, 2 ed. Brasília, DF, Pp. 416. [*Closed: correction of soil and fertilization.*]

Sousa, F. C. D., Araújo, M. P., Lemos, J. R. (2015). Ethnobotanical Study with Native Species in a Rural Village in Piauí State, Northeast Brazil. *Journal of Plant Sciences*, 3 (2), 45-53.

Sousa, H. S. de., Castro, A. A. J. F., Soares, F. A. R., Farias, R. R. S. de., Sousa, S. R. de., (2008). Florística e fitossociologia de duas áreas de cerrado do litoral, Tutóia e Paulino Neves,nordeste do Maranhão. *Publ. Avulsas Conserv. Ecossistemas,* Teresina, (21), 1-26. [Floristic and phytosociology of two cerrado coastal areas, Tutóia and Paulino Neves, northeast of Maranhão.]

Sousa, M. G., Barros, J. S., Sousa, S. R., Farias, R. R. S., Castro, A. A. J. F. (2009). Composição florística e fitossociologia das Serras de Campo Maior, município de Campo Maior, Piauí, Brasil. *Publicação Avulsas Conservação de Ecossistemas*, Teresina, (24), 1-20. [Floristic composition and phytosociology of the Sierras de Campo Maior, Campo Maior municipality, Piauí, Brazil. *Conservation of Ecosystems.*]

Tavares, S., Paiva, F. A. F., Tavares, E. J. de S., Carvalho, G. H. de., Lima, J. L. S. de. (1970). Inventário florestal de Pernambuco: I - estudo preliminar das matas remanescentes do município de Ouricuri, Bodocó, Santa Maria da Boa Vista e Petrolina. *Boletim de Recursos Naturais*, 8 (1/2), 149-194. [Forest inventory of Pernambuco: I - preliminary study of the remaining forests of the municipality of Ouricuri, Bodocó, Santa Maria da Boa Vista and Petrolina. *Natural Resources Newsletter.*]

Tavares, S., Paiva, F. A. F., Tavares, E. J. de S., Carvalho, G. H. de. (1975). *Inventário florestal na Paraíba e no Rio Grande do Norte*: *I - estudo preliminar das matas remanescentes do Vale do Piranhas.* Recife: SUDENE, Série Recursos Vegetais 4, Pp. 31. [*Forest inventory in Paraíba and Rio Grande do Norte: I - preliminary study of the remaining forests of the Piranhas Valley.]*

Tavares, S., Paiva, F. A. F., Tavares, E. J. de S., Lima, J. L. S. de. (1969a). Inventário florestal do Ceará: I - estudo preliminar das matas remanescentes do município de Quixadá. *Boletim de Recursos Naturais*, 7 (1/4), 93-111. [Forest inventory of Ceará: I - preliminary study of the remaining forests of the municipality of Quixadá. *Natural Resources Newsletter.*]

Tavares, S., Paiva, F. A. F., Tavares, E. J. De S., Lima, J. L. S. de, Carvalho, G. H. de. (1969b). Inventário florestal de Pernambuco: I - estudo preliminar das matas remanescentes do município de São José de Belmonte. *Boletim de Recursos Naturais*, 7 (1/4), 113-139. [Forest inventory of Pernambuco: I - preliminary study of the remaining forests of the municipality of São José de Belmonte. *Natural Resources Newsletter.*]

Tavares, S., Paiva, F. A. F., Tavares, E. J. De S., Lima, J. L. S. de. (1974a). Inventário florestal do Ceará: II - estudo preliminar das matas remanescentes do município de Tauá. *Boletim de Recursos Naturais*, 12 (2), 5-19. [Forest inventory of Ceará: II - preliminary study of the remaining forests of the municipality of Tauá. *Natural Resources Newsletter.*]

Tavares, S., Paiva, F. A. F., Tavares, E. J. De S., Lima, J. L. S. de. (1974b). Inventário florestal do Ceará III: - estudo preliminar das matas remanescentes do município de Barbalha. *Boletim de Recursos Naturais*, 12(2), 20-46. [Forest inventory of Ceará III: - preliminary study of the remaining forests of the municipality of Barbalha. *Natural Resources Newsletter.*]

Vasconcelos, A. D. M., Henriques, I. G. N., Souza, M. P de., Santos, W de. S., Santos, W De. S., Ramos, G. G. (2017). Caracterização florística e fitossociológica em área de Caatinga para fins de manejo florestal no município de São Francisco-PI. *ACSA*, 13 (4), 329-337. [Floristic and phytosociological characterization in the Caatinga area for purposes of forest management in the municipality of São Francisco-PI.]

Vaz, A. M. S. F., Lima, M. P. M., Marquete, R. (1992). Técnicas e manejos de coleções botânicas. In: *Manual técnico da vegetação brasileira*. IBGE, Rio de Janeiro, Manuais Técnicos em Geociências, Pp. 5-75. [Techniques and management of botanical collections. In: *Brazilian vegetation technical manual.*]

Vieira-Filho, M. A. M., Meireles, V. De. J. S., Lemos, J. R. (2018). Conhecimento popular relacionado ao uso das planas na cultura local da comunidade rural de Curral Velho. In: Lemos, J. R. (Org). *Pesquisas Botânicas e Ecológicas no Piauí*. Curitiba: CRV, Coedição: Teresina, PI: EDUFPI, Pp. 161-189. [Popular knowledge related to the use of plants in the local culture of the rural community of Curral Velho. In: *Botany and Ecological Research in Piauí*.]

Vilela, S. L. de O. (2000). *A importância das novas atividades agrícolas ante a globalização: a apicultura no Estado do Piauí*. Teresina: Embrapa Meio-Norte, Pp. 228. [*The importance of new agricultural activities in the face of globalization: beekeeping in the State of Piauí*.]

WWF-Brasil. World Wide Fund for Nature- Brasil. *Ameaças ao Cerrado*. (2018). Acesso em 7 de Fevereiro de 2018, Disponível em https://www.wwf.org.br. [*Threats to the Cerrado*.]

Zickel, C. S., Vicente, A., Almeida Jr., E. B. Cantarelli, J. R. R., Sacramento, A. C. (2004). Flora e Vegetação das Restingas do Nordeste Brasileiro. In: Eskinazi-Leça, E., Neumann-Leitão, S., Costa, M. F. *Oceanografia-Um cenário tropical*. Recife: Ed. Bagaço. Pp. 689-701. [Flora and Vegetation of Restingas of Northeast Brazil. In: *Oceanography-A tropical setting.]*

Biographical Sketches

Lucas Santos Araújo

Affiliation: Federal University of Piauí state, Brazil

Education: Undergraduate student in Biological Sciences

Research and Professional Experience: Floristics, Phytogeography and Ethnobotany.

Publications from the Last 3 Years:

http://lattes.cnpq.br/1156545441612504

Graziela de Araújo Lima

Affiliation: Autonomous professional

Education: Biologist

Research and Professional Experience: Floristics, Phytogeography and Phytosociology

Professional Appointments: Science and Biology teacher of Elementary Education

Publications from the Last 3 Years:

http://lattes.cnpq.br/9910561062003510

Jesus Rodrigues Lemos

Affiliation: Federal University of Piauí state, Brazil

Education: Doctorate in Biological Sciences – Botany

Research and Professional Experience: Floristics, Phytogeography and Ethnobotany.

Professional Appointments: Professor of Botany

Publications from the Last 3 Years:

http://lattes.cnpq.br/0603749727482775

In: Advances in Environmental Research ISBN: 978-1-53615-009-4
Editor: Justin A. Daniels

Chapter 4

THE DEMOGRAPHIC-SOCIOECONOMIC-ENTREPRENEURIAL NEXUS OF TOWNS IN A SOUTH AFRICAN BIOSPHERE RESERVE

Danie Francois Toerien*
Centre for Environmental Management,
University of the Free State, Bloemfontein, South Africa

ABSTRACT

Biosphere reserves face the challenge of sustainable development. They have to foster economic development that is ecologically and culturally sustainable. Paradoxically, the demographic-economic-entrepreneurial nexus of biosphere reserves has not been researched, an omission addressed here by studying the towns of the Gouritz Cluster Biosphere Reserve in South Africa. There is extensive orderliness in the above nexus and the interlinkages of many of its characteristics have been quantified. Systematic regularities, some time-independent, occur in the socioeconomic domain and between population sizes of towns and their enterprise numbers. Power-laws describe population-based scaling of a number of characteristics, indicating differences between small and large

* Corresponding Author Email: Toeriend@ufs.ac.za.

towns. Enterprise profiles have changed over time, with the tourism and hospitality services sector the biggest winner and the trade services sector the biggest loser. Productive knowledge, measured as enterprise richness, is a significant driver of the scaling of enterprise and population numbers. The wealth/poverty states of towns, measured as their enterprise dependency indices, modify the relationship between enterprise richness and population numbers. The development of plans for sustainable development without consideration of the quantitative orderliness of the demographic-economic-entrepreneurial nexus of this biosphere reserve and the factors that control it, would be dealing in 'a strategy of hope' rather than 'a strategy of reality'. Achieving the latter in a system with regularities, some of which that have existed over decades, is a huge challenge. Understanding of how productive knowledge can be leveraged, should be a key component in any strategy.

Keywords: biosphere reserve, sustainable development, demographic-socioeconomic-entrepreneurial nexus, enterprise profiles, enterprise richness, productive knowledge, scaling, wealth/poverty

INTRODUCTION

Human history has involved a continuing interaction between peoples' efforts to improve their well-being and the environment's ability to sustain those efforts (Clark, 1986). According to the World Commission on Environment and Development (1987), sustainable development implies meeting the needs of the present without compromising the ability of future generations to meet their own needs. There is an urgent need to integrate ecology with human demography, behavior and socioeconomics in order to understand and manage ecological patterns and processes (Liu, 2001). Designated biosphere reserves provide opportunities to study the nexus between demography and entrepreneurship in locations where sustainable development is actively pursued.

Biosphere reserves are designated by UNESCO and form part of the Man and Biosphere Program (MAB) that was launched in 1971 to provide the knowledge, skills, and human values to support harmonious relationships between people and their environment (Shelton, 1988). The aims are to achieve a combination of three complementary functions: conservation (of landscapes, ecosystems, species and genetic variation), sustainable development (fostering economic development which is ecologically and culturally sustainable), and logistical support (research, monitoring, education and training) (Pool-Stanvliet, 2013). UNESCO decided to focus on biosphere reserves as areas intended to demonstrate and develop models for balanced relationships between humans and nature (UNESCO, 1996; Bergstrand et al., 2011).

All biospheres reserves are tasked to promote sustainable growth (Pool-Stanvliet, 2013). However, conflicts often arise between the use of a region by its residents and conservation of its natural resources. This aspect has received a lot of attention in many biosphere reserves, e.g., in the Nanda Devi (Rao et al., 2000) and Khangchendzonga (Krishna et al., 2002) Biosphere Reserves in India. It is, however, surprising that detailed analyses have not yet been undertaken of resident human populations, the enterprise dynamics and their nexuses of biosphere reserves.

The Gouritz Cluster Biosphere Reserve (GCBR) in South Africa has, therefore, been selected to examine the demographic-socioeconomic-entrepreneurial relationships of a biosphere reserve. It is located in the southern Cape area of South Africa (Figure 1) and is globally unique. It is the only area in the world where three recognized biodiversity hotspots converge: Fynbos, Succulent Karoo and Maputaland-Pondoland-Albany thickets (GCBR, 2017). Two mountain ranges (the Swartberg Mountains in the north and the Langeberg/Outeniqua Mountains in the south) separate the GCBR into two separate geographic sub-regions (Figure 1). To the north and nestled between the Swartberg and the Langeberg/Outeniqua/ Tsitsikamma mountain ranges lies a semi-arid to arid valley, the Little Karoo (Burman, 1981). In the south lies a more verdant coastal plain bordered by the Langeberg mountains in the north and the Indian Ocean in the south.

Several mountain passes provide access between the two sub-regions of the GCBR; the main ones being the Tradouw Pass (connecting Swellendam and Heidelberg with Barrydale), Garcia Pass (connecting Riversdale with Ladismith) and Robinson Pass (connecting Mossel Bay with Oudtshoorn). Outeniqua Pass connects George (not part of this biosphere reserve) with Oudtshoorn (Burman, 1981).

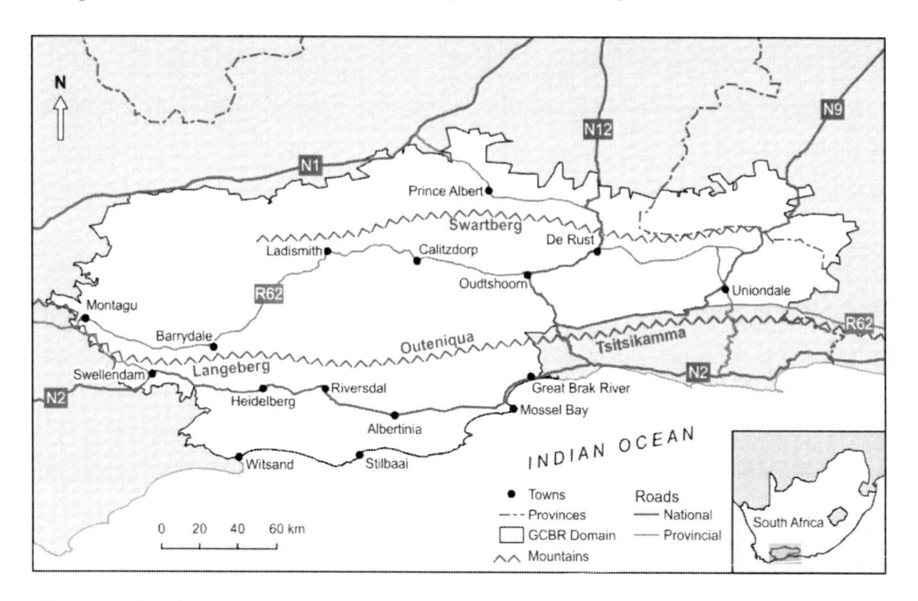

Figure 1. The Gouritz Cluster Biosphere Reserve and its towns, mountain ranges and important routes.

Some of the ecosystems of the GCBR have been degraded by previous and current practices (e.g., ostrich farming). Government-initiated programs (Working for Water, Working for Wetlands, Working for Woodlands and Working on Fire) aim to: eradicate invasive alien vegetation, facilitate and co-ordinate integrated fire management and promote and assist the restoration of degraded areas (GCBR, 2017). In order to protect and strengthen the GCBR's currently fragile ecosystems it is crucial to enhance sustainable economic development in the region. However, guidance must be based on an understanding of what development has occurred previously and how it relates to the present. In particular, the interplay of demography, socioeconomics and entrepreneurship over time in the GCBR, has to be understood.

Therefore, an investigation of the demographic and entrepreneurial nexus of the fifteen towns located in the GCBR (Figure 1) was undertaken. The towns of Montagu, Barrydale, Ladismith, Calitzdorp, Oudthoorn, De

Rust and Uniondale are located in the Little Karoo valley, which has some unique features. Burman (1981) described these as: the ostrich, a concentration of mountain passes not found elsewhere in South Africa, the Cango Caves close to Oudtshoorn, which is one of the outstanding tourist attractions in South Africa, and the tales of the struggles and successes of the people who settled in the valley. The town of Prince Albert is located on the northern slopes of the Swartberg Mountains and is linked to De Rust and Oudtshoorn.

The towns of Swellendam, Heidelberg, Riversdale, Stilbaai, Mossel Bay and Great Brak River are located on the coastal plain (Figure 1). Apart from being a strong agricultural area, a number of coastal towns (e.g., Stilbaai, Mossel Bay and Great Brak River) are holiday and retirement venues.

Several important routes cross the GCBR (Figure 1). East-west travel on the coastal plain is enabled by the N2 national road, which links Swellendam, Heidelberg, Riversdale, Albertinia, Mossel Bay and Groot Brak River. The coastal town of Stilbaai is linked to the N2 via a lesser road. East-west travel in the Little Karoo is enabled by the R62 route, which links Montagu, Barrydale, Ladismith, Calitzdorp and Oudthoorn, and which in turn is linked with De Rust and Prince Albert via the N12. Oudtshoorn is also linked with Uniondale via the N9. The N9 and N12 national routes provide north-south linkages (Figure 1). The R62 is the main roadway of Route 62, an internationally renowned tourist route in South Africa.

Rationale of this Contribution

The purpose of this contribution is to examine at a macro level the demographic-socioeconomic-entrepreneurial-wealth/poverty nexus of the GCBR and its two sub-regions (the Little Karoo and coastal plain). This examination presents historic and recent information of the nexus, which should guide efforts to overcome the challenges of the GCBR to move towards sustainable development. In developing the contribution, background information is presented first. To start, the history of the GCBR is briefly considered. This is followed by a short consideration of demographic and entrepreneurial agglomeration in human settlements. A brief review of regularities (proportionalities) in enterprise profiles and dynamics follows. Productive knowledge and the use of enterprise richness (the number of enterprise types in a town and abbreviated as ER) as its proxy are then considered. It is followed by a review of scaling phenomena in human settlements. The influences of the wealth/poverty states of towns and of historic time on the nexus are then briefly discussed. Thereafter the methods and results are presented. A discussion and conclusions complete the contribution.

A Brief History of the GCBR Area

Territories to the north and east of Table Bay, South Africa had been occupied by hunter-gatherers (San people) for millennia and by Khoikhoi herders for centuries (Elphick, 1979). The Little Karoo and coastal plain of the GCBR, thus, had a permanent human population from early times (e.g., Burman, 1981, Hopkins, 1955, Briel, 2002). In the Little Karoo, Khoikhoi inhabited the low-lying floor of the fertile valley and the San the hills (Burman, 1981). Khoikhoi tribes such as the Hessequa and Gouriqua also utilized the coastal plain (Hopkins, 1955; Briel, 2002).

The Dutch East Indies Company (VOC) founded a settlement at Table Bay in 1652 to supply its ships (Elphick, 1979). To source supplies, the VOC soon found it necessary to move beyond the Table Bay area. European thrusts into the inland areas consisted of three distinct, though overlapping, phases identified by the European agents who were most prominent in relations with the Khoikhoi: the traders, the cultivators and the pastoral stock farmers called the "trekboers" (Elphick, 1979). Burrows (1988) described some of the colonial history of the coastal plain area. By 1720 trekboers started settling to the south of the Langeberg mountains. By 1730 trekboers entered the Little Karoo. Their initial access to the Little Karoo was through gateways used by the Khoikhoi and San to cross from the coastal plain through or over the mountains: Attaquas Kloof, Platte Kloof, and Cogmans Kloof (Burman, 1981).

Due to a lack of decent roads, all trekboers in the more remote sections of the developing Cape colony were obliged to put up with excessively high transportation costs. The poorly developed transportation system and low population densities combined to prevent any significant labor specialization. Frontier residents had to forfeit many of the economic advantages and social amenities derived from a division of labor. On their isolated farms, each trekboer produced almost everything he and his family needed (Elphick, 1979) and goods such as sugar, coffee, materials and gun powder were by the 1830s mostly bartered on their farms from itinerant Jewish traders called "smouse" (Kaplan and Robertson, 1986).

After 1795 and the first British occupation of the Cape, the most obvious economic change was the development of a British trade connection and the scale of trade increased greatly (Freund, 1979). After the second British occupation in 1806 more ships visited Cape Town (De Kiewiet, 1957). Wine exports doubled between 1814 and 1824 and the export of wool transformed the Cape economy, linking it more firmly to the world market and launching it on a course of progress and relative prosperity from which enterprises in towns benefited (Tamarkin, 1996). Farmers in the Little Karoo and the Coastal Plain contributed to wool exports.

Mountain streams from the Swartberg, Langeberg and Outeniqua mountains contribute considerable run-off into the Little Karoo, assisting in the cultivation of crops (Kollenberg and Norwich, 2007). In the 19th century Oudtshoorn was already known for its tobacco production (Wickins, 1983) but its history and that of the Little Karoo is tightly linked to the story of ostrich farming (Burman, 1981). Once Oudtshoorn farmers developed methods to breed ostriches in captivity, ostrich farming became a reality, assisted by the warm dry air, plentiful water and rich food supply of the Little Karoo valley (Kollenberg and Norwich, 2007).

Ostrich feather export was only 915 kg in 1858 but by 1882 it had increased to 120 500 kg (Burman, 1981). This industry, dominated by Jewish traders (Kaplan and Robertson, 1986), brought fabulous wealth to the Little Karoo and attracted immigrants. Feather millionaires were the order of the day (Burman, 1982; Van Waart, 2001) and "ostrich barons" built fantastic 20-room mansions at Oudthoorn. After that, the ostrich industry had several boom-and-bust cycles but even at present remains an important agricultural cog in the Little Karoo economy. Farming of ostriches, however, has also contributed to the degradation of the Little Karoo ecosystem and its ecosystem services (Le Maitre et al., 2007).

Table 1. The towns of the GCBR

Sub-region	Town	Year Founded[a]	Type[a]	Population 2014	Enterprises 2014
Coastal Plain	Swellendam	1743	Administrative	17537	398
	Heidelberg	1855	Church	11586	122
	Riversdale	1837	Church	16176	283
	Albertinia	1900	Church	6887	114
	Mossel Bay	1822	Harbor	94135	1949
	Great Brak River	1886	Informal	10619	170
	Stilbaai	1894[b]	Recreational[c]	6049	255
Little Karoo	Montagu	1851	Speculative	15176	269
	Barrydale	1878	Church	4156	82
	Ladismith	1851	Church	7864	124
	Calitzdorp	1845	Church	4285	72
	Oudtshoorn	1839	Church	95933	1000
	De Rust	1900	Church	3566	54
	Prince Albert	1842	Church	7054	155
	Uniondale	1856	Church	4525	61

[a] Fransen (2006); [b] Wikipedia 2017; [c] own judgment.

Urbanization of the GCBR was slow. Only one town, Swellendam, was founded by the VOC in 1800 for administrative purposes (Fransen, 2006 and Table 1). By 1850 there were only five towns in the GCBR area (Table 1). Thereafter the pace of urbanization increased to the extent that all of the GCBR towns had been founded by 1900 (Table 1). Most towns were founded to cater for the need of farming communities for religious services and not for commercial reasons; they are the so-called 'church towns' (Fransen, 2006) (Table 1). Many of the itinerant Jewish traders became the traders in the towns and Jewish businessmen played prominent entrepreneurial roles in these towns (Kollenberg and Norwich, 2007). Apart from the church towns, there are 'recreational towns' located on the coast (Stilbaai and Great Brak River) and Mossel Bay is also a harbor town (Table 1).

Demographic and Entrepreneurial Agglomeration in Human Settlements

The way populations are distributed across geographic areas, while continuously changing, is not random (Eeckhout, 2004). There is a strong tendency toward agglomeration, i.e., the concentration of a population within common restricted areas like cities, with a few large cities and many smaller cities. A striking pattern of such agglomerations is Zipf's law for cities (Gabaix, 1999), which appears to hold in virtually all countries and dates. For instance, the size distribution of cities in the United States is startlingly well described by a simple power-law (Krugman, 1996), which essentially states that the probability that the size of a city is greater than some S, is proportional to 1/S (Gabaix, 2009). Zipf's law is a special case of a Pareto distribution (Andriani and McKelvey, 2009). The reasons for the existence of this regularity is obscure and Krugman (1996) remarked: "At this point we are in the frustrating position of having a striking empirical regularity with no good theory to account for it."

Eeckhout (2004) suggested that to provide an accurate description of agglomeration and population mobility involves accounting for the way populations are distributed over different geographic locations and accounting for their evolution over time.

Cities are a standard unit of observation in urban economics, just as countries are a norm in international economics (Rose, 2006). A project early in the new millennium at the Santa Fe Institute in the U.S. investigated if the same sort of analyses used to understand biological network systems, could be used for studying cities and companies (West, 2017). They soon reported the presence of remarkable proportionalities in the characteristics of many cities. For instance, they recorded that power-laws described the distribution of many characteristics and that some properties of cities scale with population size (West, 2017). The scaling exponents fall in distinct universality classes. Quantities reflecting wealth creation and innovation have exponents larger than unity (showing increasing returns) whereas those accounting for infrastructure have exponents smaller than unity, showing economies of scale (Bettencourt et al., 2007a). The Santa Fe research showed that, despite appearances, cities are approximately scaled versions of one another. The extraordinary regularities open a window on studying the underlying mechanisms, dynamics and structure common to all cities (Bettencourt and West, 2010). Cities are remarkably robust: success, once achieved, is sustained for several decades or longer, thereby setting a city on a long run of creativity and prosperity (Bettencourt and West, 2010).

The reason why some cities become large is essentially because of inertia in the creation of jobs (Gabaix, 1999). The number of new jobs is roughly proportional to the number of existing jobs. Agglomeration and residential mobility of the population between different geographic locations are tightly connected to economic activity (Eeckhout, 2004). The evolution of populations across geographic locations is an extremely complex amalgam of incentives and actions taken by many individuals, businesses, and organizations. Economic factors are the principal determinant of the dynamics of city populations (Eeckhout, 2004).

The demographic and enterprise distributions of the GCBR towns were included in a study of demographic distributions of regional towns in South Africa (Toerien, 2018a). Pareto-like population and enterprise distributions close to Zipf's law apply to towns of different South African regions, including the GCBR towns. This was true for two time periods (1946/47 and 2013/14) about 67 years apart. The research of the Santa Fe and South African groups has clearly shown that there are demographic-entrepreneurial regularities in many urban settlements in the world. Given the challenge of pursuing sustainable development in the GCBR, it is necessary to understand how the demographic-socioeconomic-entrepreneurial nexus of the GCBR has changed over time. The time-independent demographic and entrepreneurial orderliness implied by these results must be considered in connection to the sustainable development challenge of the GCBR.

Demographic-Socioeconomic-Entrepreneurial Dynamics of Cities

Social, Economic and Other Interdependent Facets of Cities

Cities as systems, are complex with many interdependent facets, e.g., social, economic, infrastructural, and spatial characteristics. There are surprisingly systematic regularities and similarities in the demographic and socioeconomic nexus of cities (e.g., Bettencourt et al., 2007a; Bettencourt et al., 2007b; Bettencourt et al., 2010; Bettencourt & West, 2010; West, 2017). Cities are more than the linear sum of their individual components (Bettencourt et al., 2010). These regularities open a window onto underlying mechanisms, dynamics, and structures common to all cities and strongly suggest that all of these phenomena are in fact highly correlated and interconnected, driven by the same underlying dynamics and constrained by the same set of "universal" principles that enable a conceptual framework to deal with the problems caused by urbanization (West, 2017) or the challenges stemming from the pursuit of sustainable development.

Toerien (2015a) reported on a range of interrelated demographic-economic-entrepreneurial characteristics of local municipalities in the Free State province of South Africa. These included: gross value added (GVA), population numbers, employment numbers and enterprise numbers. He concluded that there is a strong underlying structure in the economic, demographic, and entrepreneurial domains of these municipalities, which is probably part of a system in which economic value addition, population size, employment creation, and entrepreneurial spaces are linked to one another. This structure is probably largely driven by the magnitude of money in local economies.

The challenges involved in pursuing sustainable development in biosphere reserves should consider the complexities outlined above.

Enterprise Proportionalities, Profiles and Dynamics

Academic studies of South African towns started in the 1960s (e.g., Davies, 1967; Davies and Cook, 1968; Davies and Young, 1969). Three broad strands of enquiry about small towns in South Africa have been later been discerned: (i) small town growth and development potential related to a core theme of local economic development (LED) (e.g., Hoogendoorn and Nel, 2012), (ii) examination of various small town activities relative to a debate on post-productivist landscapes (e.g., Hoogendoorn and Nel, 2012; Hoogendoorn & Visser, 2016), and, (iii) studies of the regularities (proportionalities) in enterprise development and the demographics of towns (e.g., Toerien and Seaman, 2010, 2011, 2012a, 2012b).

A study of 125 South African towns reported extensive regularities (Toerien and Seaman, 2012a). These proportionalities were in the form of statistically significant correlations between entrepreneurial (e.g., total number of enterprises in towns or number of enterprises in certain business sectors in towns), economic (e.g., gross regional value added, total regional personal income and regional employment) and demographic (e.g., town populations) characteristics (Toerien, 2014a, 2015a, 2015b; Toerien and Seaman, 2010; 2011; 2012a, 2012b, 2012c, 2012d; 2014). In towns of the Eastern Cape Karoo of South Africa, statistically significant correlations between population numbers and enterprise numbers were present during a period of about a century (1911 to 2006) (Toerien, 2014b). The observed correlations suggest that the 'total entrepreneurial spaces' of towns are probably governed by the size of their populations and that most, if not all, of the entrepreneurial spaces are fully occupied by entrepreneurs.

Two broad entrepreneurial types were detected in these towns: 'run-of-the-mill' entrepreneurs and 'special' entrepreneurs. This induces differences in enterprise development dynamics (Toerien and Seaman, 2012a). 'Run-of-the-mill' enterprises are dependent on, and are limited by, local demand. Moretti (2013) reported that the vast majority of jobs in a modern society are in local services, e.g., people that work as waiters, plumbers, nurses, teachers, etc. They offer services that are produced and consumed locally, in other words, in the non-traded sector. This sector seems to be similar to the 'run-of-the-mill' sector of Toerien and Seaman (2102a).

Moretti (2013) added that most of the jobs in innovative industries belong to the traded sector, together with jobs in traditional manufacturing, some services, and the agricultural and extractive industries. They produce tradable goods or services that are mostly sold outside a region. The 'special' entrepreneurs of Toerien and Seaman (2012a) are probably active in producing or delivering tradeable products and services. The paradox is that while the vast majority of jobs are in the non-traded sector, the traded sector is the driver of prosperity in the U.S. cities (Moretti, 2013). This could also be generally true for biosphere reserves.

Sustainable development in biosphere reserves must by definition involve the founding of new sustainable enterprises. It is necessary to examine the enterprise profiles and dynamics of the GCBR and its sub-regions to determine to what extent, if any, this challenge has already been met.

Enterprise Richness and Productive Knowledge

Youn et al. (2016) determined the relative abundance of business types as a function of U.S. city sizes. This sheds further light on the processes of economic differentiation in human settlements. In South Africa, the ER of towns has been examined (Toerien and Seaman, 2014; Toerien, 2017). ER refers to the total number of enterprise types present in a town. It is a measure of business diversity and is analogous to the species richness concept used in ecological studies (Spellerberg and Fedor, 2003). A statistically significant log-log (power-law) relationship was recorded between the total enterprise numbers and ER in a large group of South African towns (Toerien and Seaman, 2014), including GCBR towns (Toerien, 2017). It was also shown that the power-law relationship has endured virtually unchanged over approximately 70 years (Toerien, 2017). The fact that towns with more enterprises than others always also have more enterprise types, is a clear indication that the economic growth that results in the expansion of enterprise numbers in towns, small or large, is partly dependent on new business ideas and the founding of enterprises based on enterprise types not yet present in (i.e., new to) growing towns. As the size of towns increases, new enterprise types are needed in a constant log-log ratio to total enterprises.

Schumpeter (1942) referred to the process of industrial mutation that incessantly creates new businesses and Hausmann and Klinger (2006) argued that producing new things is quite different from producing more of the same. Florida (2002) remarked: "Human creativity is the ultimate economic resource. The ability to come up with new ideas and better ways of doing things is ultimately what raises productivity and thus living standards." The entrepreneurial wellbeing of towns is clearly linked to the ability to conceive and deliver new products and/or services, i.e., having more productive knowledge.

Wealth and development are related to the complexity that emerges from the interactions between the increasing number of individual activities that constitute an economy (Hidalgo and Hausmann, 2009). Hausmann et al. (2017) produced an atlas of economic complexity based on data extracted from 128 countries representing 99% of world trade. They concluded that the differential accumulation of productive knowledge distinguishes between rich and poor countries. These differences are expressed in the diversity and sophistication of the things that each of these nations makes. The productive knowledge to create new products or services is key to economic success and wealth. It is more than book knowledge or searches on the Internet. It is embedded in brains and human networks and is tacit and hard to transmit and acquire. It comes more from years of experience than from years of schooling (Hausmann et al., 2017).

Toerien (2018b; Toerien, 2018c) argued that the entrepreneurial wellbeing of South African towns is clearly connected to the ability of their residents to conceive and make new products and/or deliver new services. In other words, their entrepreneurial wellbeing is dependent on the level of their productive knowledge and ER can be used as a surrogate measurement of their productive knowledge. The concepts of ER and productive knowledge provide new ways to examine the socioeconomic dynamics of South African towns (Toerien, 2018b; 2018c).

Sustainable development in biosphere reserves by definition involves the start-up of enterprise types not yet present and requires additional productive knowledge. It is, therefore, important to quantify the productive knowledge present in biosphere reserves and to relate it to the demographic-socioeconomic-entrepreneurial nexuses of biospheres. Being able to determine the ERs of towns in the GCBR provides the means of doing that for this biosphere reserve.

Scaling as an Important Issue

With urban populations increasing dramatically worldwide, cities are playing an increasingly critical role in human societies and in sustainability (Bettencourt et al., 2010). Linear per capita indicators are generally used to characterize and rank cities. These indicators implicitly ignore the fundamental role of non-linear agglomeration in the life history of cities. Non-linear population agglomeration is explicitly manifested by the super-linear power law scaling of most urban socioeconomic indicators with population size (Bettencourt, 2013; Ortman et al., 2014). As a result, larger cities are disproportionally the centers of innovation, wealth and crime, and all to approximately the same degree (Bettencourt et al., 2010). Local urban dynamics display long-term memory, so a city that under or outperforms its size expectation maintains such (dis)advantage for decades.

The characteristics of modern cities take a simple mathematical form. Based on population, N(t), as the measure of city size at time t, the power-law scaling is:

$$Y(t) = Y_0 N(t)^{\beta} \qquad (1)$$

Y can denote material resources (such as energy or infrastructure) or measures of social activity (such as wealth, patents, and pollution); Y_0 is a normalization constant. The exponent, β, reflects the general dynamic rules at play across urban systems.

Robust and commensurate scaling exponents have been recorded across different nations, economic systems, levels of development, and recent time periods for a wide variety of indicators (Bettencourt et al., 2007a). Measures of the physical extent of urban infrastructure increase more slowly than city population size, thus exhibiting economies of scale. They scale sub-linearly ($\beta < 1$). Regardless of where a city is located and regardless of the specific metric used, only about 85% more material infrastructure is needed for every doubling of city populations (West, 2017). On the other hand, various socioeconomic outputs increase faster than population size and thus exhibit increasing returns to scale. They scale super-linearly ($\beta > 1$), which is typical of open-ended complex systems (Ortman et al., 2015; West, 2017). Large cities are environments where there are more sustained social interactions per unit time. These generic dynamics, in turn, are the basis for expanding economic and political organization, such as the division and coordination of labor, the specialization of knowledge, and the development of political and civic institutions (Ortmann et al., 2015). These dynamics provide the basis for cities to be the gateways for ideas (Glaeser, 2011).

South African towns have been analyzed on a per capita basis (e.g., Toerien and Seaman, 2012a, 2012b, 2012c). Scaling impacts of agglomeration in South African towns have not been examined before. It is, therefore, important to determine if scaling occurs as a potentially important factor in the demographic-entrepreneurial nexus of the GCBR towns.

Wealth/Poverty as a Regulating Factor

Poverty is relative and the poor are lacking the resources with which to attain a socially acceptable quality of life (May, 2012). A range of regularities, which have been interpreted in terms of entrepreneurship, have been recorded in the enterprise dynamics of South African towns (e.g., Toerien and Seaman, 2010, 2011, 2012a, 2012b, 2012c, 2012d). Two of the statistically significant correlations of the enterprise numbers of South African towns are: (i) a linear relationship between enterprise numbers and population numbers, and, (ii) a log-log relationship (power-law) between enterprise numbers and ER (as a surrogate for productive knowledge). To unravel the impact of poverty on entrepreneurial development in South African towns, Toerien (2018b; 2018c) used these two relationships to link the population numbers, productive knowledge (as ER) and EDI (the Enterprise Dependency Index, a wealth/poverty measure) of towns in South Africa.

From a theoretical viewpoint, linear regressions between the population and total enterprise numbers in South African towns can be stated as:

$$\text{Enterprises} = \text{b(Population)} + \text{C} \quad (2)$$

The regression coefficient b is:

$$\text{b} = \text{Enterprises/Population} \quad (3)$$

Toerien (2014b) suggested that the inverse of this regression coefficient, (i.e., 1/b), which relates to how many persons are associated with the average enterprise in a group of towns that exhibit such a linear correlation, is a measure of the wealth/poverty state of the group of towns because it measures the number of persons needed to 'carry' the average enterprise in the group. Therefore:

$$1/\text{b} = \text{EDI} = \text{Population numbers/Enterprise numbers} \quad (4)$$

More persons per enterprise in a town indicates more poverty, and fewer persons per enterprise, wealthier conditions.

Equation 3 can be restated as:

$$\text{Enterprise numbers} = \text{Population numbers/EDI} \quad (5)$$

The enterprises in South African towns are thus related to the size of their populations as well as their wealth/poverty states. EDI can, therefore, be used to examine the influence of wealth/poverty states of GCBR towns on their demographic-entrepreneurial nexuses.

Toerien (2018b, 2018c) provides quantitative proof of the contention of Hausmann et al. (2017) that productive knowledge and wealth/poverty are linked. Toerien (2018c) reported that the poverty state of South African towns (measured as EDIs) has a severe influence on enterprise dynamics, both in terms of total enterprises as well as on the numbers of enterprise types. For instance, a rich town with EDI < 80 and with an initial population of 50 000 residents, will have 718 total enterprises and 210 enterprise types. A poorer town with a similar population size and EDI >300, will have only 79 total enterprises and 47 enterprise types. After 5 years at a growth rate of 2 percent per annum, the former town will have 774 total enterprises and 222 enterprise types and the latter town only 85 enterprises and 49 enterprise types. The entrepreneurial challenges and job creation abilities of poorer towns are clearly different from those of wealthier towns. The influence of the poverty/wealth states of GCBR towns should be considered in terms of their development challenges.

Time as a Regulating Factor

Fransen (2006) traced the beginnings and early development of towns and villages in South Africa. Before 1795, the monopolistic VOC frowned upon private enterprise in the few towns of its settlement. For a long time, travelling peddlers served the requirements of farmers, the majority of the rural population (Tamarkin, 1996). In 1819, the British government, which formally annexed the Cape in 1815, decided to encourage immigration to the Cape Colony. The European population increased from just over 26 000 in 1806 to about 110 000 in 1855. It had become necessary to develop a network of service centers in the Cape Colony, then an area about the size of the United Kingdom. These services were mostly religious and/or administrative in nature. The commercial element only became important during the course of the 19th century (Tamarkin, 1996).

The preponderance of parish establishments as a reason for town foundation in the Cape Colony (Fransen, 2006) was probably higher than in most other parts of the Dutch and English colonial empires (Tamarkin, 1996). Historical accounts indicate that successful entrepreneurs developed significant rural business empires in the 1800s in the Cape Colony, e.g., the Mosenthal brothers who linked their Port Elizabeth business with branches in Graaff-Reinet, Richmond, Murraysburg, Burgersdorp, Hopetown and Aliwal North (Tamarkin, 1996) and Joseph Barry who created a business empire in the southern Cape region (Burrows, 1988). Quantified information about enterprise development in the early towns is, however, extremely limited. For instance, Ross recorded that in 1865 Graaff-Reinet had eleven bakers, 26 boot and shoe makers, 35 blacksmiths and 108 wholesale and retail merchants (Tamarkin, 1996).

Toerien and Seaman (2010) developed methods based on examination of telephone directories to identify, quantify and classify enterprises in South African towns. The methods were also applied to a telephone directory of 1946/47 and revealed that regularities similar to those recently registered (e.g., Toerien and Seaman, 2012a) were present in South African towns in 1946/47 (Toerien and Seaman, 2014; Toerien, 2017). It is, therefore, possible to examine the evolution over time of the demographic-entrepreneurial nexus of GCBR towns.

Methods

Demographic Data

Demographic data for 1904 to 1970 of all GCBR towns was sourced from official census data (Republic of South Africa, 1976). Demographic data for 2001, 2006 and 2011 was sourced from City Population (no date) (a website accessed at https://www.citypopulation.de). The population estimates for 2005/06 were based on 2001 data extended by the growth rates of towns between 2001 and 2011. The population estimates for 2013/14 were based on 2011 data extended by the growth rates of towns between 2001 and 2011.

Socioeconomic Data

Socioeconomic data such as gross value added (GVA), total income and employment numbers are not freely available for towns in South Africa but at the lowest level only for magisterial districts. Therefore, such information for the magisterial districts present in the GCBR were purchased from the firm Global Insight, which extracted data for 2011 from the databank of Statistics South Africa. The magisterial districts included: Calitzdorp, Heidelberg, Ladismith, Montagu, Mossel Bay, Oudtshoorn, Prince Albert, Riversdale, Swellendam and Uniondale. Ordinary least squares (OLS) analysess were used to determine the interrelationships between: GVA, total income, employment numbers and enterprise numbers (determined in this study). Microsoft Excel software was used for the analyses.

Enterprise Data, Profiles and Dynamics

Enterprise data of each town was compared for three periods many years apart: 1946/47, 2005/06 and 2013/14. The following telephone directories were used to obtain the information: 1946/47 (Cape Times Ltd, 1946), 2005/06 (Trudon, 2005) and 2013/14 (Trudon, 2013).

The enterprises of the GCBR towns were identified, classified into 19 different business sectors and enumerated according to Toerien and Seaman (2010). This procedure yielded the enterprise profile of each town and for each time period. This information was used to to develop an enterprise profile for each of the sub-regions and the GCBR in total for each of the years. Where appropriate Microsoft Excel software was used for correlation and regression analyses to elucidate the demography-enterprise-poverty relationships of GCBR towns.

To enable sensible comparisons of the enterprise profiles of the GCBR and its sub-regions, the enterprise numbers of each business sector for a specific year were normalized by division by the total number of enterprises for that year and expressed as a percentage. To enable sensible comparisons of the economic strengths of each business sector and for each year, the enterprise profiles of the GCBR and its sub-regions were also normalized by expression of the number of enterprises per thousand residents.

For scaling analyses, the data values of each pair of characteristics being investigated were normalized through division by their averages. Thereafter log-log regression analyses were done to determine if statistically significant power-laws were present, an if so, if sub-linear, linear, or super-linear scaling (West, 2017) was present. Microsoft Excel software was used for the analyses.

Enterprise Richness (ER)

The enterprise type of each of the enterprises of the GCBR towns was determined from a database of more than 700 enterprise types encountered in South African towns (Toerien and Seaman, 2014; Toerien, 2017). That provided the ER of each town for each time period.

The linear relationship between population and enterprise numbers (see equation 2) and the power-law between ER and the population numbers of each town for each time period were determined as described above. Microsoft Excel software was also used for these calculations.

Toerien (2017) reported that the ER-enterprise numbers power-law for South African towns is enduring. To determine if this is also the case in the GCBR towns, the power-law relationship between the enterprise numbers and the ER-values of all towns was determined for the total time period.

To quantify the impact of increases of ER, the impacts of a single ER-unit added to different-sized GCBR towns have been calculated. In 2014 the populations of the GCBR towns varied from about 3600 (De Rust) to about 96000 (Oudtshoorn) (Table 1) and their ER-values ranged from 24

(De Rust) to 276 (Mossel Bay). Use was made of a highly significant ER-enterprise numbers power-law (to be presented later) to calculate enterprise numbers of towns for four levels of ER-values: 30, 100, 200 and 300. Employment numbers were then calculated using an ordinary linear regression ratio of employees per enterprise (to be presented later) to calculate the employment numbers of the four levels. To assess the impact of the addition of a single ER-unit the same set of calculations were done for another four ER levels, namely 31, 101, 201 and 301, and the increases in enterprise and employment numbers due to the increase in ER-value calculated.

Enterprise Dependency Index (EDI) and Wealth/Poverty as Regulating Factor

The number of persons per enterprise is a measure of how many people are needed to 'carry' the average enterprise of a town or a region. As explained earlier the number of enterprises in a town/municipality is a function of the population size of the town/municipality and the wealth/poverty status of its population (see equation 5).

For individual towns, EDI is derived from a simple calculation of the population number divided by the enterprise number. For regions and sub-regions, it is derived from the inverse of the regression coefficient obtained from a regression analysis of population numbers of towns of the region (or sub-region) (as independent variables) and enterprise numbers (as dependent variables) of the same towns in the region (or sub-region). The inverse of the regression coefficient represents the number of persons per average town in the region (or sub-region).

The regulating influence of the wealth/poverty state of towns on the demographic-entrepreneurial nexus of the GCBR towns was investigated by: (i) using groups of towns binned according to the levels of their EDIs. This allowed investigations of three groups of towns: 1. wealthier, 2. intermediate, and 3. poorer towns, and, (ii) relating the wealth/poverty states of the towns to growth scenarios.

Personal Interviews

Because of the success of the tourism and hospitality sector in both sub-regions of the GCBR, personal interviews were done with persons central to the establishment of Route 62 in the Little Karoo, now an internationally renowned South African tourist route.

RESULTS

Systematic Socioeconomic Regularities

The GCBR municipalities are complex with many interdependent facets: demography, value addition, total personal income, employment and entrepreneurial development (Table 2).

Table 2. The systematic socioeconomic regularities of the ten magisterial districts of the Gouritz Cluster Biosphere Reserve as determined by ordinary least squares (OLS) analyses. Per capita data is for 2011

Independent Variable	Dependent Variable	Correlation*	R^2	Slope	Intercept	n
Employment	Income	0.96	0.927	0.187	-270.3	10
Enterprises	Employment	0.89	0.794	10.47	4563.7	10
GVA	Employment	0.95	0.906	3.80	2666.5	10
GVA	Enterprises	0.98	0.953	0.332	-122.6	10
GVA	Income	0.98	0.966	0.761	130.5	10
GVA	Population	0.92	0.846	14.98	8831.2	10
Income	Enterprises	0.97	0.947	0.427	-165.8	10
Income	Population	0.94	0.890	19.84	6014.5	10
Population	Employment	0.98	0.952	0.239	957.4	10
Population	Enterprises	0.88	0.768	0.018	-175.6	10

* All correlations are statistically significant ($P < 0.01$). GVA = gross value added. R^2 = variation explained.

These realities, here expressed on a per capita basis, have to be considered in the pursuit of sustainable economic development in the GCBR. Changes in one characteristic are likely to result in changes in others.

Demography and Entrepreneurial Growth in the GCBR

Population and Enterprise Numbers

The population and the enterprise numbers of the GCBR as well as it sub-regions have increased substantially since 1946 (Table 3). Over time, the population and enterprise growth of the coastal plain outstripped that of the Little Karoo, perhaps reflecting the general trend of people migrating to coastal areas (e.g., Stimson and Minnery, 1998). Between 1946 and 2014, population growth was higher than enterprise growth in the whole of the GCBR, reflecting a slow overall increase in the general poverty level. Population growth and enterprise growth between 1946 and 2014 were more evenly balanced in the coastal plain than in the Little Karoo. The latter probably reflects the decline of the ostrich industry as a primary driver of economic growth, which led to an increase of poverty in the Little Karoo. However, there is a need to better understand the insidious impacts of poverty in the GCBR and its sub-regions, something reported on later in this contribution. Between 2006 and 2014 the population and enterprise growth ratios were fairly similar in the Little Karoo and the coastal plain, perhaps indicating reasonably even growth dynamics. Population growth and enterprise growth are now proceeding hand-in-hand.

Linear Relationships

There were statistically significant ($P < 0.01$) linear relationships over the 1946 to 2014 period between population numbers and enterprise numbers in the GCBR and its sub-regions (Table 3). In excess of 74% of the variation is explained in all cases (see R^2 in Table 3). As would be expected, towns with larger populations have more enterprises and smaller towns fewer. However, what is surprising is that strong proportionality between population numbers and enterprise numbers occurred in all three time periods. Youn et al. (2016) reported a similar proportionality in U.S. cities. The proportionalities enabled estimates of the EDI-values of the GCBR and its sub-regions (Table 4). The EDI-values of the Little Karoo have increased between 1946 and 2006, indicating rising poverty. That of the coastal plain has decreased markedly, which indicates rising prosperity in the sub-region.

Table 3. The population and enterprise numbers for the period 1946 to 2014 of towns in the Gouritz Cluster Biosphere Reserve (GCBR), its Little Karoo sub-region and its coastal plain sub-region

Year & ratio	Population			Enterprises		
	GCBR	Little Karoo	Coastal Plain	GCBR	Little Karoo	Coastal Plain
1946	54430	30776	23654	992	504	488
2005/06	264714	129731	134983	4146	1493	2653
2013/14	305548	142559	162989	5108	1817	3291
46:06	4.9	4.2	5.7	4.2	3.0	5.4
46:14	5.6	4.6	6.9	5.1	3.6	6.7
06:14	1.2	1.1	1.2	1.2	1.2	1.2

46:06, 46: 14 and 06:14 reflect the ratios of numbers between 1946, 2006 and 2014 respectively.

Table 4. The linear relationships between population and enterprise numbers in the Gouritz Cluster Biosphere Reserve (GCBR) and its sub-regions. All correlations are statistically significant ($P < 0.01$)

Year	Region	Correlation	R^2	Regression coefficient	Intercept	EDI
1946/47	GCBR	0.94	0.877	0.0131	18.5	76.2
2005/06		0.93	0.859	0.0132	44.1	76.0
2013/14		0.92	0.854	0.0152	31.2	65.8
1946/47	Little Karoo	0.98	0.968	0.0126	14.5	79.2
2005/06		0.99	0.988	0.0096	40.2	104.7
2013/14		0.99	0.988	0.0100	48.7	99.9
1946/47	Coastal Plain	0.87	0.748	0.0160	15.8	62.6
2005/06		0.99	0.984	0.0181	29.5	55.2
2013/14		0.99	0.986	0.0208	-13.5	48.1

Enterprise Profiles and Changes over Time

The total enterprises as well as enterprises in 19 different business sectors of the towns of the GCBR and its sub-regions are presented in Table 5. The table also reflects the growth in total enterprises between 1946 and 2014. This data was used to obtain the enterprise profiles of the GCBR and its sub-regions. The enterprises of each sector are expressed as a percentage of the total enterprises. To visualize the changes over time, the 1946 value of each sector was subtracted from the 2014 value for the same sector and the resulting data were ranked from smallest to largest. Figure 2 presents the changes for the GCBR.

Table 5. The total as well as business sector enterprises of the GCBR and its sub-regions, the Little Karoo and the coastal plain for 1946/47, 2005/06 and 2013/14

	1946/47			**2005/06**			**2013/14**		
Business sector	GCBR	Little Karoo	Coastal Plain	GCBR	Little Karoo	Coastal Plain	GCBR	Little Karoo	Coastal Plain
Agricultural Products & Services	41	25	16	166	69	97	180	74	106
Processors	62	38	24	84	45	39	91	43	48
Factories	49	22	27	21	5	16	31	8	23
Construction Services	76	30	46	368	81	287	405	113	292
Mining Services	1	0	1	6	0	6	5	1	4
Tourism & Hospitality	71	29	42	613	279	334	1374	567	807
Engineering & Technical Services	29	14	15	89	26	63	100	26	74
Financial Services	54	28	26	237	94	143	233	94	139
Legal Services	35	23	12	68	24	44	70	20	50
Telecommunication Services	1	1	0	33	11	22	43	10	33

	1946/47			2005/06			2013/14		
Business sector	GCBR	Little Karoo	Coastal Plain	GCBR	Little Karoo	Coastal Plain	GCBR	Little Karoo	Coastal Plain
News & Advertising Services	9	5	4	8	3	5	15	6	9
Trade Services	273	146	127	1010	390	620	961	348	613
Vehicle Services	48	26	22	302	98	204	320	112	208
General Services	64	26	38	237	74	163	295	71	224
Professional Services	10	5	5	122	46	76	112	35	77
Personal Services	71	39	32	260	95	165	310	101	209
Health Services	52	28	24	260	90	170	309	119	190
Transport & Earthworks	46	19	27	101	28	73	97	29	68
Real Estate Services	0	0	0	160	35	125	157	40	117
Total Enterprises	992	504	488	4145	1493	2652	5108	1817	3291

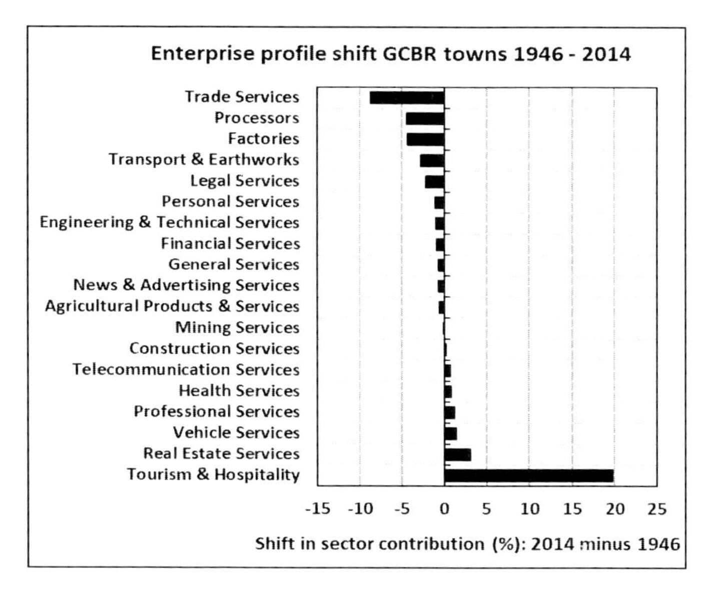

Figure 2. The shifts in the enterprise profile of the GCBR towns between 1946 and 2014.

The tourism sector had grown the most of all sectors between 1946 and 2014, followed by real estate services and professional services. The sectors that have decreased the most are trade services, processors (i.e., enterprises that add value to local primary products), factories (i.e., enterprises that add value to materials sources external to a specific local economy) and legal services.

The sector providing tourism and hospitality services is now numerically the largest business sector in the GCBR. Natural and man-made attractions form the basis of a large part of this sector. Therefore, its growth probably reflects growth in sustainable development. From a sustainability point of view, however, it is disconcerting that the processor and factory sectors have weakened. In addition, the losses in trade services probably reflect the replacement of small general dealers by large national retail chains offering products that were produced locally in earlier times.

The overall number of enterprises per thousand capita has remained remarkably similar from 1946 to 2014 (Figure 3). There was a strong decline in the strength of the trade sector of both sub-regions leading to the same trend overall in the GCBR (Figure 4). The decline in the trade sector was largely offset by the strong growth of the tourism and hospitality sector in both sub-regions and the GCBR (Figure 5).

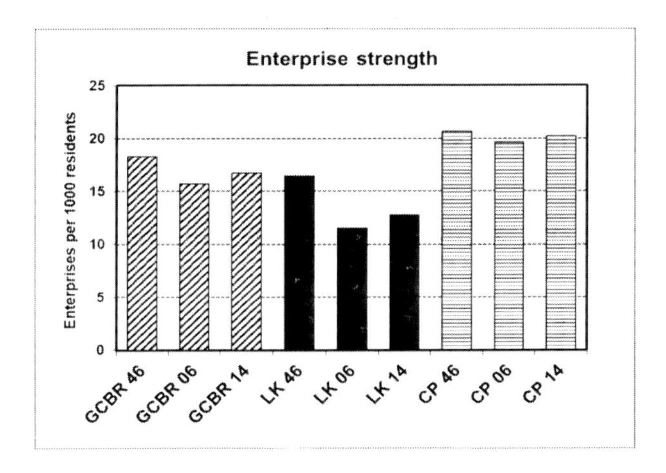

Figure 3. The time-dependency of enterprises per thousand capita of the Gouritz Cluster Biosphere Reserve (GCBR) and its sub-regions, the Little Karoo (LK) and the coastal plain (CP). 46 = 1946; 06 = 2006 and 14 = 2014.

The Tourism and Hospitality Sector

The two largest towns of the GCBR house the largest number of tourism & hospitality enterprises (Figure 6). Mossel Bay, the largest, is on the coastal plain, and Oudtshoorn, the second largest, is in the Little Karoo. Tourism is important in both sub-regions (as is also shown in Figure 5). However, there is an issue that needs explanation.

With exception of Stilbaai, all of the towns of the coastal plain are located along the N2 national road that leads from Cape Town to the Garden Route (Figure 1), one of South Africa's best-known tourist destinations. A comparison of the tourism and hospitality strength (expressed as enterprises per thousand residents) of all of the GCBR towns (Figure 7), however, shows that the towns of the Little Karoo have generally stronger tourism and hospitality sectors than the towns on the coastal plain despite the latter ones being on a national route on the way to a famous tourist destination. Because of its implications for sustainable development, the issue was investigated further.

Personal interviews indicated that two persons, Gert Lubbe, owner of a hotel in the town of Montagu, and Jeanetta Marais, then CEO of the Breede River Valley District Council, cooperated in the development of Route 62, a then new tourist route through the Little Karoo. Lubbe needed to get more people to visit his hotel and Ms Marais wanted to promote tourism in the region. Based on the realization that Cape Town and the Garden Route are two of the most visited tourism destinations in South Africa, Lubbe argued that Montagu should be linked to this reality. Together with Ms Marais who provided expert knowledge and funding for tourism development, the pair undertook the task of establishing Route 62 that links the towns of the Little Karoo to the Garden Route and other tourist destinations. Effective promotion of Route 62 to tour operators then created one of the present premier tourist routes in South Africa and the Little Karoo towns benefited. For instance, Montagu has at present the fourth highest number of tourism and hospitality enterprises in the GCBR (Figure 6). All towns of the Little Karoo have benefited from the realization of the vision developed by a couple of far-sighted people, which created sustainable economic development based on the scenic beauty and other attractions of the Little Karoo.

On the coastal plain, tourism is mainly linked with coastal and marine experiences (Mossel Bay and Stilbaai) and with history (e.g., the historic town of Swellendam) (Figures 6 and 7). These aspects should also be considered for further sustainable development.

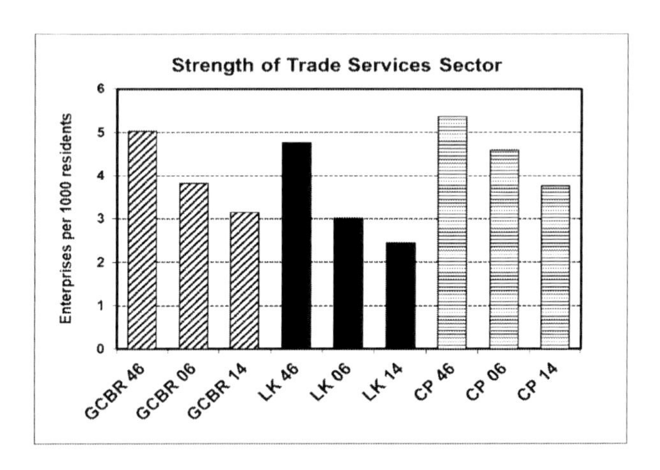

Figure 4. The time dependency of the trade services enterprises per thousand capita of the Gouritz Cluster Biosphere Reserve (GCBR) and its sub-regions, the Little Karoo (LK) and the coastal plain (CP). 46 = 1946; 06 = 2006 and 14 = 2014.

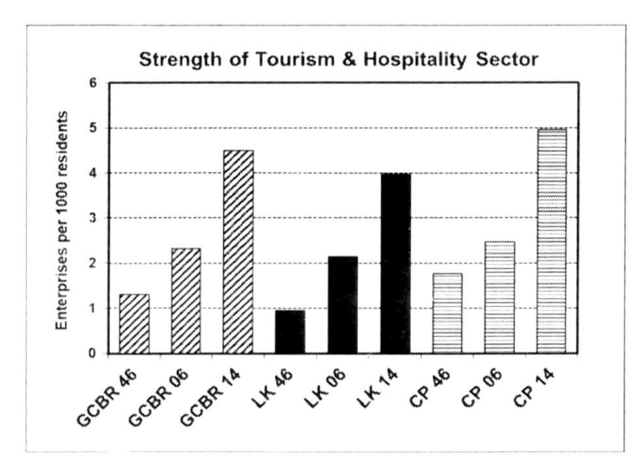

Figure 5. The time-dependency of the tourism and hospitality services enterprises per thousand capita of the Gouritz Cluster Biosphere Reserve (GCBR) and its sub-regions, the Little Karoo (LK) and the coastal plain (CP). 46 = 1946; 06 = 2006 and 14 = 2014.

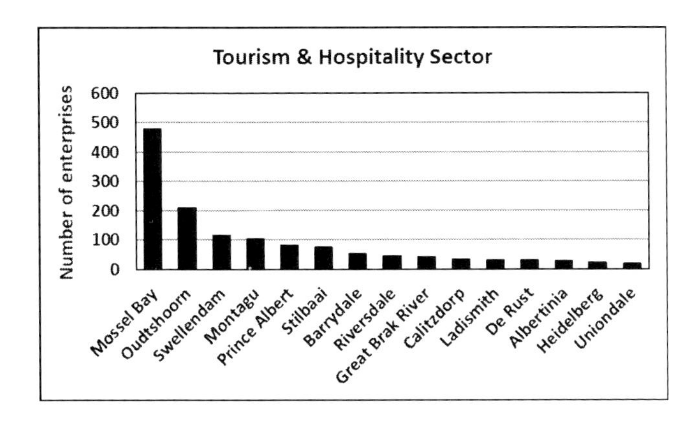

Figure 6. The number of tourism and hospitality enterprises in of the towns of the Gouritz Cluster Biosphere Reserve (GCBR).

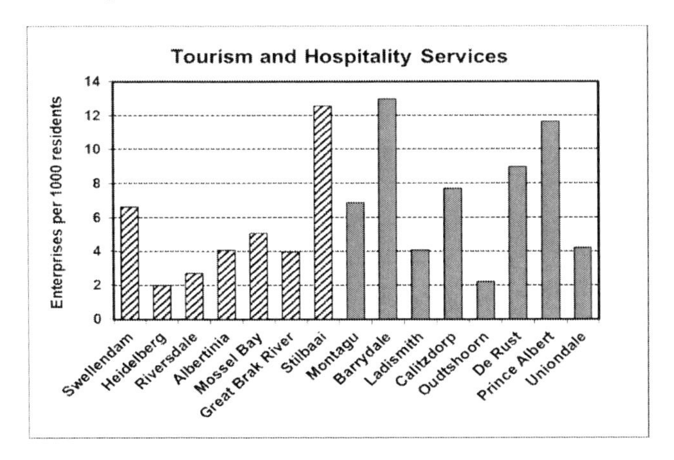

Figure 7. The strength of the tourism and hospitality sector in the towns of the GCBR (measured as enterprises per 1000 residents). The striped blocks indicate towns on or linked to the N2 national road passing through the coastal plain and the grey blocks indicate towns on Route 62 in the Little Karoo.

Lock-in

Lock-in is a situation in which a region's or country's economy becomes over-reliant on or dominated by one of more economic sectors (Martin and Sunley, 2010). To determine if lock-in has occurred in the entrepreneurial development of the GCBR, the correlations between the enterprise profiles of the GCBR for 1946, 2006 and 2014 and each of the profiles of the sub-regions for the different years were determined (Table

6). The closer the years are to each other, the higher the correlations. All of the correlations are statistically significant (P < 0.01) indicating that the enterprise structures changed only slowly over the years, which suggests that some lock-in occurred. This is another issue to consider for further sustainable development in the GCBR.

Table 6. Correlation coefficients between the enterprise profiles of Gouritz Cluster Biosphere (GCBR) (independent variable) and those of its sub-regions (LK = Little Karoo and CP = coastal plain) (dependent variables)

GCBR 46	LK 46	0.99	**GCBR 46**	CP 46	0.99
	LK 06	0.86		CP 06	0.88
	LK 14	0.59		CP 14	0.67
GCBR 06	LK 46	0.85	**GCBR 06**	CP 46	0.90
	LK 06	0.98		CP 06	0.99
	LK 14	0.84		CP 14	0.90
GCBR 14	LK 46	0.58	**GCBR 14**	CP 46	0.69
	LK 06	0.92		CP 06	0.84
	LK 14	0.99		CP 14	1.00

46 = 1946; 06 = 2006; 14 = 2014.

Enterprise Richness and Productive Knowledge

The Enterprises-ER Relationship

A statistically significant power-law with sub-linear exponent of 0.7017 describes the relationship between enterprise numbers and ER in the GCBR towns for all of the time periods (Figure 8). The power-law has endured over more than 65 years and illustrates that enterprise development and dynamics in GCBR towns are subject to an orderliness, which cannot be changed by whims. ‘Strategies of hope’ for sustainable development, often based on the misconception that there are ‘silver bullets’ that can be achieved by little tweaks of the system, will not work. Effective strategies should be based on the realities of the orderliness that has existed for many decades. For instance, two of the implications of the enduring relationship (Figure 8) are: (i) a town with double the enterprises of another town will have about 63% more enterprise types than the other town, and, (ii) a town with half the enterprises of another town, will have some 61% fewer enterprise types.

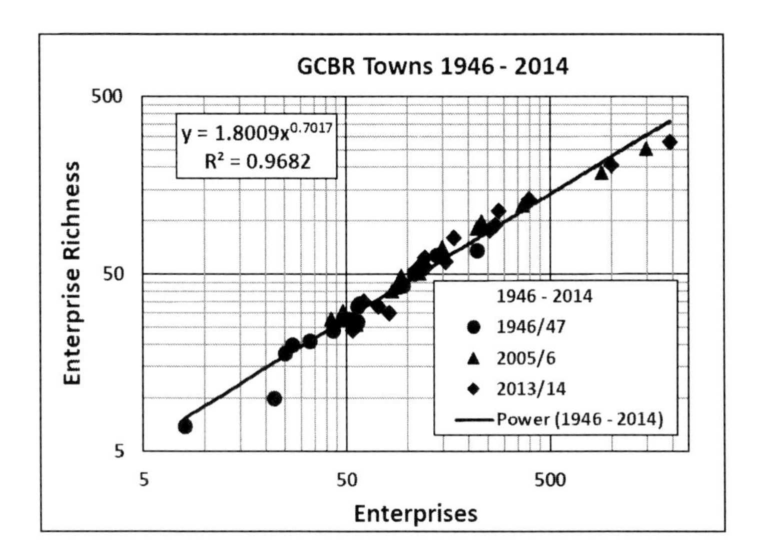

Figure 8. The enduring power-law that describes the relationship between enterprise numbers and enterprise richness (ER) in GCBR towns.

The power-law of Figure 8 was used to examine the dynamics of what Toerien (2015b, 2017) labeled as 'new' and 'existing' entrepreneurship in towns. New entrepreneurship is the ability to conceive and successfully implement a new business idea in a local economy/town, e.g., being the first medical practitioner in a town that has not had any medical practitioners to date. The number of times new enterprise types have been successfully started in a town is reflected in the ER-value of that town. Existing entrepreneurship is the start-up of an enterprise of a type already present in a local economy/town (e.g., a second legal firm). The proportional representation of these two entrepreneurial types relative to the entrepreneurial spaces of GCBR towns (measured by total enterprise numbers), are presented in Figure 9. In villages and small towns of the GCBR with fewer than 80 to 90 enterprises, there is a need for more 'new entrepreneurship' than 'existing entrepreneurship'. This represents a significant challenge for small communities with a few enterprises. Beyond 80 to 90 enterprises, the need for 'existing entrepreneurship' grows steadily, reaching an 80:20 pareto distribution by about 1200 enterprises. Importantly, economic growth of large towns always includes a need for 'new' entrepreneurship, e.g., about 20% in towns with 1200 enterprises. The ability to conceive and implement new business ideas is an absolute requirement, which should be considered in the challenge of sustainable development in biosphere reserves.

ER-Based Scaling in GCBR Towns

West and colleagues have investigated scaling impacts linked to the population dynamics of cities (West, 2017). Their rationale was the fact that the large population growth expected in the world over the next decades as well as the continuing of migration of rural people to cities, will result in numerous new mega-cities. There is, therefore, a need to understand population size as a driver in the socioeconomic dynamics of cities. The link between ER and enterprises (Figure 8) adds a question about the role of ER (i.e., productive knowledge) in scaling impacts.

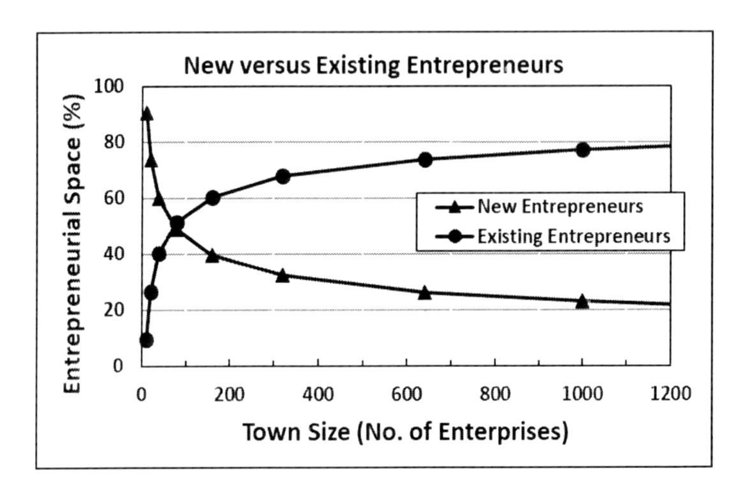

Figure 9. The proportional needs for new and existing entrepreneurship in different-sized GCBR towns.

Inverting the power-law of Figure 8, means that enterprise numbers are described as a function of ER (i.e., productive knowledge). This relationship (Figure 10) is also very tight (96% of the variation is explained), statistically significant ($P < 0.01$) and has lasted for at least 65 years. Enterprise numbers scale super-linearly and strongly as a result of increases in ER, i.e., productive knowledge.

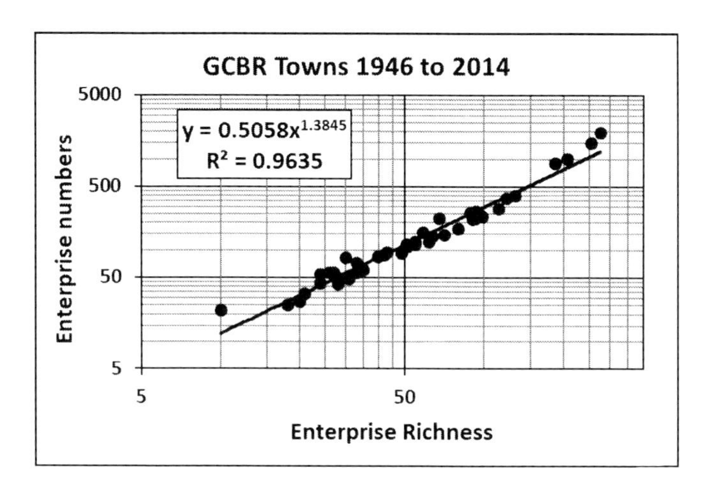

Figure 10. The GCBR power-law describing the time-independent log-log relationship between enterprise richness (ER) and enterprise numbers.

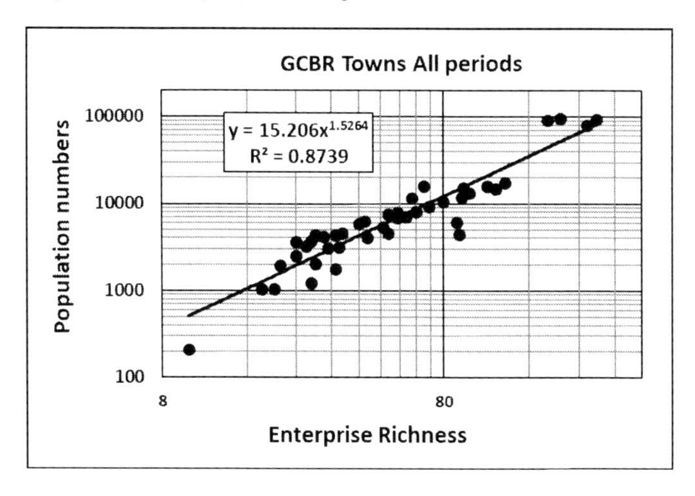

Figure 11. The GCBR power-law between enterprise richness (ER) and population numbers over all periods.

In the GCBR enterprise numbers are correlated with population numbers (Table 2) and ER (productive knowledge) is correlated with enterprise numbers (Figure 10). It follows that ER and population numbers should be correlated, and this is the case for all time periods (Figure 11). The exponent of this power-law of 1.53 is super-linear, indicating returns to scale, i.e., towns with more productive knowledge have proportionally more people (this evidence of scaling is examined further later on).

The Influence of Wealth/Poverty

The data points of Figure 11 are more widely spread than those of the relationship between ER and enterprises (see Figure 10). Toerien (2018b) showed that the wealth/poverty states of towns of the Eastern Cape Karoo, South Africa alter the relationship between ER-level and population numbers. To investigate if this is also the case in the GCBR, the influence of the wealth/poverty states (measured by EDI) of GCBR towns on their ER-population size relationship was examined (Figure 12).

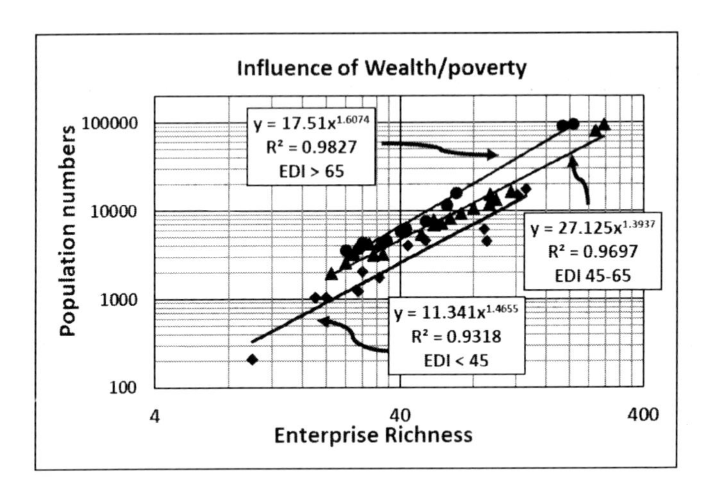

Figure 12. The power-laws between enterprise richness (ER) and population numbers over all periods for three groups of GCBR towns binned on basis of their wealth/poverty states: diamonds, EDI <45 – wealthier group; triangles, EDI 45-65 – intermediate group; dots, EDI > 65 - poorer group.

The wealth/poverty states of towns indeed influence the ER-population relationship. Increasing poverty moves the power law describing this relationship upwards in Figure 12. A specific level of productive knowledge is tightly related to a specific number of enterprises (Figure 10). However, the same level of productive knowledge can be associated with more than one level of population numbers as a result of differences in the wealth/poverty state of towns (Figure 12). The wealth/poverty states of GCBR towns are important in sustainable development considerations.

Scaling Phenomena in GCBR Towns

The impacts of scaling as a result of the agglomeration of people in cities have been investigated extensively by West and colleagues (West, 2017). Agglomeration occurs in GCBR towns and enterprise and population numbers scale with increases in productive knowledge (ER) (Figures 10 and 11 respectively. Therefore, scaling associated with population numbers as well as productive knowledge was investigated further in the GCBR towns.

Scaling Based on Population Numbers

To examine population-based scaling in the GCBR, use was made of the socioeconomic data for 2014 obtained from Global Insight. Three statistically significant power-laws were recorded:

The relationship between population size and GVA:

$$\text{GVA (Rand million)} = 0.9215(\text{population})^{1.1977} \tag{6}$$

has a super-linear exponent indicating returns to scale, i.e., proportionally more value is added in towns with larger populations than towns with smaller populations.

The relationship between population size and total personal income:

$$\text{Total personal income (Rand million)} = 0.7864(\text{population})^{1.369} \tag{7}$$

also has a super-linear exponent indicating returns to scale, i.e., proportionally more total personal income is earned in larger towns than smaller towns. The relationship between population size and employment:

$$\text{Employment} = 1.0114(\text{population})^{0.9908} \quad (8)$$

has a linear exponent that indicates that employment and populations increase/decrease in step with each other.

The relationship between population size and number of enterprises:

$$\text{Enterprises (no.)} = 1.0095(\text{population})^{0.8291} \quad (9)$$

has a sub-linear exponent that indicates an economy to scale similar to what has been registered for infrastructural characteristics in U.S. cities (e.g., Bettencourt et al., 2010; West, 2017). Larger towns have proportionally fewer enterprises than smaller towns.

Agglomeration of people in GCBR towns results in similar scaling impacts as those registered for U.S. and other cities (e.g., Bettencourt, 2013; Bettencourt et al., 2007a, Bettencourt et al., 2007b, Bettencourt et al., 2010, Bettencourt & West, 2010, West, 2017). GCBR towns are also more than the linear sum of their individual components and this issue should be considered in dealing with the challenges of sustainable development.

Scaling Based on Productive Knowledge (ER)

Enterprise numbers (Figure 10) and population numbers (Figure 11) are both functions of productive knowledge (measured as ER). Both power-laws have super-linear exponents that are larger than 1.5. Increases in productive knowledge (ER) drives strong super-linear scaling in population and enterprise numbers in the GCBR. The potential impacts of such scaling on additional enterprise and employment numbers (method described earlier) are presented in Table 7.

In a small town with an ER-value of 30 and some 90 enterprises (somewhat similar to Calitzdorp, De Rust and Uniondale), the addition of a new enterprise type should result in a total increase of five enterprises (the new enterprise plus four more of types already present). The gain in terms of employment should be about 50 jobs (Table 7). In a large town with an ER-value of 300 and just over 3000 enterprises (somewhat similar to Mossel Bay and Oudtshoorn), the gain would be 16 enterprises (the new business type plus 15 other enterprises of types already in the town) and an increase of some 160 new jobs.

Table 7. The potential impact of a single new enterprise type (one ER-unit) on opportunities for additional enterprises and employment in GCBR towns

Enterprise richness	Associated enterprises (no.)	Increase in enterprises*	Potential employment created
30	56		
31	59	3	27
100	297		
101	301	4	43
200	776		
201	781	5	56
300	1360		
301	1366	6	66

* includes the new enterprise type.

The remarkable value of productive knowledge (Hausmann et al., 2017) and the significant leverage of enterprises that produce tradeable products (Moretti, 2013) are also available to GCBR towns. Increasing productive knowledge is a key to grow the economic sustainability of the GCBR and to increase sustainable job opportunities.

The Moderating Influence of Wealth/Poverty

Figure 12 depicts the influence of the wealth/poverty states of GCBR towns (measured as EDI) on their ER-population numbers power-law relationships and Figure 10 depicts the power-law relationship between ER and enterprise numbers of these towns. These power-laws are used here to illustrate how increasing poverty necessitates increasingly larger populations to carry a specific number of enterprises (Table 8). The ratio of the richer towns to the poorer towns in terms of population needed to sustain the enterprises also expands with increasing town sizes. It increases from 2.6 for smaller towns to 3.6 for larger towns. Both poverty levels and town sizes are important and should be considered for sustainable economic development of biospheres.

Table 8. Illustration of how increasingly larger populations are needed to carry a specific number of enterprises related to a specific level of productive knowledge (ER) with increasing poverty levels (measured as enterprise dependency index, EDI). Ratio = population of poorer towns (EDI > 65) to population of richer towns (EDI < 45)

Enterprise Richness	Enterprises	Population needed to carry enterprises			Ratio
		EDI < 45	EDI 45-65	EDI > 65	
40	84	2526	4636	6583	2.61
80	218	6976	12182	20059	2.88
120	382	12638	21435	38491	3.05
200	776	26718	43683	87491	3.27
300	1360	48403	76866	167885	3.47
400	2025	73785	114779	266586	3.61

DISCUSSION

Despite the fact that one of the challenges of biosphere reserves is to enable sustainable development (Pool-Stanvliet, 2013), the demographic-entrepreneurial-wealth/poverty nexus of biosphere reserves has not been investigated. This contribution addresses that omission and uses the GCBR, a fascinating and unique biosphere reserve in South Africa, as a case study to show that there are many reasons why such studies are necessary.

Cities/Towns and Regularities

Research in the U.S. has shown that cities are approximately scaled versions of one another and there are surprisingly systematic regularities and similarities in the demographic, socioeconomic and infrastructural nexus of cities. These extraordinary regularities open a window on studying the underlying mechanisms, dynamics and structures common to all cities (e.g., Bettencourt et al., 2007a; Bettencourt et al., 2007b; Bettencourt et al., 2010; Bettencourt and West, 2010; West, 2017). The GCBR municipalities are also complex open systems with many interdependent facets (Tables 2, 4, 6, Figures 8, 10, 11, 12). The demographic and socioeconomic results are similar to those reported for municipalities in the Free State (Toerien, 2015a) and are in line with the suggestion of Toerien and Seaman (2012d) that value addition (measured as GVA) is an important driver of other socioeconomic characteristics of South African towns. These findings should be considered in the handling of the challenges of sustainable economic development.

In the GCBR and over time, the population and enterprise growth of the coastal plain outstripped that of the Little Karoo (Table 3), perhaps reflecting the general trend in the world of people migrating to coastal areas (e.g., Stimson and Minnery, 1998). There was also strong proportionality between population numbers and enterprise numbers in 1946/47, 2005/06 and 2013/14 (Table 3) and such proportionality is probably a permanent characteristic of the GCBR towns. Youn et al. (2016) reported a similar proportionality for U.S. cities. These proportionalities indicate that there are limits to what can be achieved socio-economically in GCBR towns and the limits should be heeded in development plans.

The economies of the coastal towns (measured as enterprises per thousand residents) are somewhat stronger than those of the towns of the Little Karoo (Figure 3), probably as a result of the aforementioned migration of people to coastal towns and a decline in the traditional agricultural strength due to ostrich farming in the Little Karoo. It is not evident if sustainable development could reverse the latter trend.

There was some evidence of economic lock-in over time (Table 6), which would complicate the challenge to achieve sustainable development. The major changes between 1946/47 and 2013/14 in the enterprise profiles of GCBR towns were identified (Figure 2). The tourism and hospitality sector gained the most ground (Figure 5) and the trade services sector lost the most (Figure 4). The GCBR towns have also lost some of their ability to add value to local primary produce or externally sourced materials and this has resulted in fewer processors and factories in the towns (Figure 2). In the middle of the previous century, towns in the GCBR were dependent on: (i) the abilities of dressmakers, tailors, shoemakers etc. to produce goods such as clothes, shoes, etc., and, (ii) many small, and often family-owned, general dealers, supplied groceries and other goods. Now retail chains provide these goods and some productive knowledge has been lost by the GCBR.

The Rise of Tourism

The growth of the tourism and hospitality sector in both sub-regions of the GCBR (Figure 5) provides the strongest evidence of sustainable development based on the natural resources in the GCBR, i.e., the scenic beauty of the Little Karoo and the coastal and marine resources of its coastal plain. It was possible to identify that two far-sighted persons with a shared vision have developed a successful strategy to initiate and promote Route 62, a then new tourist route through the Little Karoo. Their efforts made a significant contribution to sustainable development in the Little Karoo and show that individuals can make a difference. This sets the scene for the further development of tourism in the GCBR. In this regard the archaeological assets of the region (e.g., Marean et al., 2007) perhaps offer special potential.

Productive Knowledge and Scaling

The differential accumulation of productive knowledge distinguishes between rich and poor countries (Hausmann et al., 2017). Should this also be true for towns there are two questions: (i) how to measure productive knowledge in towns, and, (ii) how to measure the wealth/poverty states of towns. Toerien (2018b, 2018c) used ER to measure productive knowledge in towns and EDI to measure the wealth/poverty states of towns. This approach was also used successfully in this contribution.

ER refers to the total number of enterprise types present in a town (Toerien and Seaman, 2014; Toerien, 2017). Toerien (2018a, 2018b, 2018c) used ER as a surrogate measure of the productive knowledge in towns, a practice followed in this contribution. The power-law relationship between ER, i.e., productive knowledge, and enterprise numbers in GCBR has stayed the same for more than 65 years (Figure 10). Toerien (2017) also reported on the enduring nature of this power-law. The power-law must be considered as a permanent characteristic of GCBR towns.

The implications of the power-law should be heeded. For instance, the expansion of enterprise numbers in GCBR towns is partly dependent on new business ideas. The growth of any GCBR town requires the successful founding of enterprise types not yet present in the town (Figure 9). Plans for sustainable development have to consider how the productive knowledge of the GCBR or its sub-regions can be enhanced in ways compatible with the challenges facing biosphere reserves.

Scaling as a result of the agglomeration of people is a very important characteristic of cities (e.g., Bettencourt et al., 2007a, Bettencourt et al., 2007b, Bettencourt et al., 2010, Bettencourt & West, 2010, West, 2017). Many of the power-laws that describe such scaling behavior have exponents in the order of 1.15, indicating super-linear returns to scale. Evidence is presented here of scaling phenomena in the demographic-socioeconomic-entrepreneurial nexus of GCBR towns. GVA and total personal income scale super-linearly with population numbers (equations 6 and 7). Employment scales linearly (equation 8) and enterprise numbers scale sub-linearly (equation 9) with population numbers. Toerien (2018a, 2018b) reported two power-law relationships: (i) between ER and enterprises, and (ii) between ER and populations. The same was true in this contribution (Figures 10 and 11). Both of these relationships have major implications for the pursuit of sustainable development in the GCBR.

The ER-enterprises relationship has a super-linear exponent of 1.38 (Figure 10), which indicates that enterprises scale with increasing levels of productive knowledge. Every doubling of ER (= doubling of productive knowledge) results in 2.6 times more enterprises. Because of the proportionality between enterprises and employment (Table 2), significant leverage of employment opportunities should follow the expansion of enterprise numbers; an important consideration in the pursuit of sustainable development.

Most of the jobs in innovative industries belong to the traded sector, together with jobs in traditional manufacturing, some services, and the agricultural and extractive industries (Moretti, 2013). These industries produce tradeable goods or services that are mostly sold externally to a region. The entrepreneurial wellbeing of South African towns is clearly connected to the ability of their residents to conceive and make new products and/or deliver new services (Toerien, 2018b, 2018c). 'Special' entrepreneurship (Figure 9) could probably play an important role in the production of tradeable goods or provision of services for external markets.

The ER-population number relationship (Figure 11) has a super-linear exponent of 1.53 which indicates that population numbers also scale with increasing productive knowledge. Every doubling of productive knowledge in a town, results in 2.9 times more people in the town. Larger towns carry proportionally more people than smaller towns. Town-size needs to be considered in sustainable development plans.

Should the relationship between ER and population numbers be generally true for all cities, the results presented here quantify the significant leverage potential of productive knowledge suggested by Hausmann et al. (2017). Population scales super-linearly with ER and many other characteristics scale super-linearly with population (Bettencourt et al., 2010; Lobo et al., 2013). Expansion of ER, thus, produces a 'double-whammy': more productive knowledge leads to proportionally more people, proportionally more value addition and proportionally more employment. These impacts should be heeded in the pursuit of the sustainable development goal of biospheres.

The Influences of Wealth/Poverty

To unravel the impact of poverty on entrepreneurial development in South African towns, Toerien (2018b, 2018c) demonstrated that the number of enterprises in a town is a function of the population size of the town and the wealth/poverty state of the town (measured as EDI). Smaller EDIs indicate more wealth, larger EDIs more poverty. This study has confirmed the EDI influences the relationship between the population numbers and enterprise numbers of GCBR towns. Poverty has increased in the Little Karoo towns between 1946 and 2006 but there was rising prosperity in the towns of the coastal plain (Table 4). Binning of the GCBR towns into three groups (richer, intermediate and poorer) and examination of the ER-population power-laws of the three groups (Figure 12) indicated that the wealth/poverty states of the three groups altered the characteristics of the power-laws. Wealth/poverty states of towns clearly matter and the entrepreneurial challenges of poorer towns are clearly different from those of wealthier towns. The influence of the poverty/wealth states of GCBR towns must be considered when dealing with their development challenges.

Conclusion

The combination of three complementary functions are the responsibility of biosphere reserves: conservation, sustainable development, and logistical support (Bergstrand et al., 2011). The second function implies that development in the world has to some extent been unsustainable and should be reconsidered. This function of biosphere reserves, therefore, requires changes in the way development is handled; in other words, not more of the same. The comparative norm, should, therefore be the quantitative measurement of what has been done before. The primary goal of this contribution was to supply quantified information about the demographic-economic-entrepreneurial nexus of GCBR towns.

Pool-Stanvliet (2013) described sustainable development as the fostering of economic development that is ecologically and culturally sustainable. The fact that economic sustainability is not mentioned, might indicate a 'biosphere paradigm' in which the demographic-socioeconomic-entrepreneurial nexus of biosphere reserves is not perceived to be important. This contribution has illustrated the contrary.

The official or entrepreneur in the GCBR, and perhaps those in other biosphere reserves too, whose goal it is to expand the number of sustainable enterprises in his/her biosphere reserve, faces an exceedingly difficult challenge. To be successful, he/she must expand the number of sustainable enterprises in a system that has an extensive orderliness in its demographic-economic-entrepreneurial nexus and offers limited local opportunities. The development of plans for sustainable development without consideration of the quantitative information about the nexus and the factors that control it, deals in 'strategies of hope' rather than 'strategies of reality'. Bettencourt and West (2010) warned that the difference between 'policy as usual' and policy led by a new quantitative understanding of cities may well be the choice between creating a 'planet of slums' or finally achieving a sustainable, creative, prosperous, urbanized world expressing the best of the human spirit. This warning also applies to development plans for biosphere reserves.

ACKNOWLEDGMENTS

Funding to cover the basic expenses of the first part of the study was provided by the Board of Directors of the Gouritz Cluster Biosphere Reserve. The Centre for Environmental Management, University of the Free State provided general support. Frank Sokolic prepared the map of the Gouritz Cluster Biosphere Reserve. Marie Toerien and Estelle Zeelie provided technical support.

REFERENCES

Andriani, P. and McKelvey, B. (2009). From Gaussian to Paretian thinking: Causes and implications of power laws in organizations. *Organization Science*, 20(6): 1053-1071.

Bergstrand, B., Björk, F. and Molnar, S. (2011). *Biosphere entrepreneurship. Report of: Biosphere Reserve Lake Vänern Archipelago and Mount Kinnekulle, Kinnekulle, Sweden*. Accessed at: http://media.vanerkulle.org/2013/09/297_Biosphere-Entrepreneurship-A-Pilot-Study-Webversion.pdf.

Bettencourt, L. M. A. (2013). The origins of scaling in cities. *Science*, 340: 1438-1441.

Bettencourt, L. M. A., Lobo, J.., Helbing, D, Kühnert, C. and West, G. B. (2007a). Growth, innovation, scaling, and the pace of life in cities. *PNAS,* 104(17): 7301–7306.

Bettencourt L. M. A., Lobo, J. and Strumsky, D. (2007b). Invention in the city: Increasing returns to patenting as a scaling function of metropolitan size. *Research Policy* 36: 107–120.

Bettencourt, L. M. A., Lobo, J., Strumsky, D. and West, G. B. (2010). Urban scaling and its deviations: Revealing the structure of wealth, innovation and crime across cities. *PLoS ONE,* 5(11): e13541. doi:10.1371/journal.pone.0013541.

Bettencourt, L. and West, G. (2010). A unified theory of urban living. *Nature,* 467: 912-913.

Briel, R. (2002). Albertinia: Honderd jaar - 1902-2002. In *Afrikaans* [Albertinia: Hundred years, 1902-2002]. Paarl: Lourette.

Burman, J. (1981). *The Little Karoo.* Cape Town: Human & Rousseau.

Burrows, E. H. (1988). *Overberg outspan: A chronicle of people and places in the South Western districts of the Cape*. Swellendam: Swellendam Trust.

Cape Times, 1946. South African Directory 1946-47. Cape Town: Cape Times Ltd.

City Population (no date). *Population statistics for countries, administrative areas, cities and agglomerations*. Accessed at: https://www.citypopulation.de/.

Clark, W. C. (1986). In: W. C. Clark and R. E. Munn (Eds.) *Sustainable development of the biosphere*. Cambridge: Cambridge University Press, pp. 5-48.

Davies, R. J. (1967). The South African urban hierarchy. *South African Geographical Journal,* 49(1):9–19.

Davies, R. J. and Cook, G. P. (1968). Reappraisal of the South African urban hierarchy. *South African Geographical Journal,* 50(2):116–32.

Davies, R. J. and B. S. Young. 1969. The economic structure of South African cities. *South African Geographical Journal,* 51(1):19–37.

De Kiewiet, C. W. (1957). *A history of South Africa: Social and economic.* London: Oxford University Press.

Eeckhout, J. (2004). Gibrat's law for (all) cities. *American Economic Review,* 2004;94(5):1429–1451. https://doi.org/10.1257/00028280430 52303.

Elphick, R. (1979). The Khoisan to c. 1770. In: R. Elphick and H. Giliomee (Eds.), *The shaping of South African society,* 1652-1820. Maskew Miller Longman: Cape Town. Pp. 3–40.

Florida, R. (2002). *The rise of the creative class.* New York: Basic Books.

Fransen, H. (2006). *Old towns and villages of the Cape: A survey of the origin and development of towns, villages and hamlets at the Cape of Good Hope.* Johannesburg: Jonathan Ball.

Freund, W. M. (1979). The Cape under the transitional government, 1795-1814. In: R. Elphick and H. Giliomee (Eds.), *The shaping of South African society, 1652-1820.* Maskew Miller Longman, Cape Town. Pp. 211-240.

Gabaix, X. (1999). Zipf's law for cities: An explanation. *Quarterly Journal of Economy,* 114(3):739–767. https://doi.org/10.1162/00335539955 6133.

Gabaix, X. (2009). Power laws in economics and finance. *Annual. Review of Economy,* 1:255–293.

Glaeser, E. (2011). *Triumph of the city: How our greatest invention makes us richer, smarter, greener, healthier, and happier.* New York, Penguin Press.

Gouritz Cluster Biosphere Reserve (GCBR), (2017). *Our biodiversity.* Accessed at: https://www.gouritz.com/our-biodiversity/.

Hausmann, R., Hidalgo, C. A., Bustos, S., Coscia, M., Chung, S., Jimenez, J., Simoes, A. and Yıldırım, M. A. (2017). *The Atlas of economic complexity: Mapping paths to prosperity.* Report of: Center for International Development, Harvard University: Cambridge.

Hausmann, R., Klinger, B. (2008). South Africa's export predicament. *Economics of Transition*, 16(4): 609-637.

Hidalgo, C. and Hausmann, R. (2009). The building blocks of economic complexity. *PNAS,* 106 (26): 10570-10575.

Hoogendoorn, G. and Nel, E. (2012). Exploring small town development dynamics in rural South Africa's post-productivist landscapes. In: R, Donaldson and L. Marais (Eds.) *Small town geographies in Africa: Experiences from South Africa and elsewhere*. New York: Nova Science Publishers, pp. 21-34.

Hoogendoorn, G. and Visser, G. (2016.) South Africa's small towns: A review on recent research. *Local Economy*, 31(1): 95-108.

Hopkins, H. C. (1955). Eeufees-gedenkboek van die Ned. Geref. Kerk, Heidelberg (Kaapland): 1855-1955. In *Afrikaans:* [Centurion memorial of the Dutch Reformed Church, Heidelberg (Cape Province)]. Stellenbosch: Pro Ecclesia.

Kaplan, M. and Robertson, M. (1986). *Jewish roots in the South African economy.* Cape Town: Struik Publishers.

Kollenberg, A. and Norwich, R. (2007). *Jewish life in the South African country communities. Vol. III. Johannesburg: The South African Friends of Beth Hatefutsoth.* Pp. 230-242.

Krishna, A. P., Chhetri, S. and Singh, K. K. (2002). Human dimensions of conservation in the Khangchendzonga Biosphere Reserve: The need for conflict prevention. *Mountain Research and Development*: 22 (4): 328-331.

Krugman, P. (1996). Confronting the mystery of urban hierarchy. *Journal of Japanese and International Economies,* 10(4):399–418. https://doi.org/10.1006/jjie.1996.0023.

Le Maitre, D. C., Milton, S. J., Jarvain, C., Colvin, C. A. Saayman, I. and Vlok, J. H. J. (2007). Linking ecosystem services and water resources: landscape-scale hydrology of the Little Karoo. *Frontiers in Ecology and the Environment*, 5(5): 261–270.

Liu, J. (2001). Integrating ecology with human demography, behavior, and socioeconomics: Needs and approaches. *Ecological Modelling,* 140: 1–8.

Lobo, J., Bettencourt, L. M. A., Strumsky, D. and West, G. B. (2013). Urban scaling and the production function for cities. *PLoS ONE* 8(3): e58407. doi:10.1371/journal.pone.0058407.

May, J. (2012). Smoke and mirrors? The science of poverty measurement and its application. 2012. *Development Southern Africa* 29(1): 63-75.

Marean, C. W., Bar-Matthews, M., J., Bernatchez, J., Fisher, E., Goldberg, P., Herries, I. R., Jacobs, Z., Jerardino, A., Karkanas P., Minichillo T., Nilssen, P. J., Thompson E., Watts, I and Williams, H. M. (2007). Early human use of marine resources and pigment in South Africa during the Middle Pleistocene. *Nature,* 449: 905-908. doi:10.1038/nature06204.

Martin, R. and Sunley, P. (2010). The place of path dependence in an evolutionary perspective on the economic landscape. In: R. Boschma and R. Martin (Eds.). *The handbook of evolutionary economic geography.* Cheltenham: Edward Elgar, pp. 62-92.

Moretti, E. (2013). *The new geography of jobs*. New York, Mariner Books.

Ortman, S. G., Cabaniss, A. H. F., Sturm, J. O. and Bettencourt, L. M. A. (2014). The pre-history of urban scaling. *PLoS ONE*, 9(2): e87902. doi:10.1371/journal.pone.0087902.

Ortman, S. G., Cabaniss, A. H. F., Sturm, J. O. and Bettencourt, L. M. A. (2015). Settlement scaling and increasing returns in an ancient society. *Sci. Adv.* 1: e1400066, pp8.

Pool-Stanvliet, R. (2013). A history of the UNESCO Man and the Biosphere Programme in South Africa. *South African Journal of*

Science, 109(9/10): Art. #a0035, 6 pages. http://dx.doi.org/10.1590/sajs.2013/a0035.

Rao, K. S., Nautiyal, S., Maikhuri, R. K. and Saxena, K. G. (2000). Management conflicts in the Nanda Devi Biosphere Reserve, India. *Mountain Research and Development*, 20, (4): 320-323.

Republic of South Africa (1976). *Population of South Africa 1904 – 1970.* Report no. 02-05-12, Republic of South Africa. Pretoria: Government Printer.

Rose, A. K. (2006). Cities and countries. *Journal of Money, Credit and Banking,* 38(8): 2225-2245.

Schumpeter, J. A. (1942). *Capitalism, socialism and democracy*. Third Ed. New York: Harper Colophon.

Shelton, N. (1988). Ecosystem redevelopment. *Ambio,* 17 (2): 155-157.

Spellerberg, I. F. and Fedor, P. J. (2003). A tribute to Claude Shannon (1916–2001) and a plea for more rigorous use of species richness, species diversity and the 'Shannon–Wiener' Index. *Global Ecology and Biogeography,* 12: 177–179.

Stimson, R. J. and Minnery, J. (1998). Why people move to the 'sun-belt': A case study of long-distance migration to the Gold Coast, Australia. *Urban Studies* 35(2): 193-214.

Tamarkin, M. (1996). *Cecil Rhodes and the Cape Afrikaners: The imperial colossus and the colonial parish pump*. Jeppestown: Jonathan Ball.

Toerien, D. F. (2012). Enterprise proportionalities in the tourism sector of South African towns. In: M. Kasimoglu (Ed.), *Visions of global tourism industry: Creating and sustaining competitive strategies*. Rijeka: Intech. pp. 113-138. http://dx.doi.org/10.5772/37319.

Toerien, D. F. (2014a). The enterprise architecture of Free State Towns. *Technical Report*, DTK. Accessed at: https://www.researchgate.net/profile/Daan_Toerien/publications.

Toerien, D. F. (2014b). 'n Eeu van orde in sakeondernemings in dorpe van die Oos-Kaapse Karoo. LitNet Akademies, 11(1): 330-371. In *Afrikaans* [A century of order in enterprises of towns of the Eastern Cape Karoo].

Toerien, D. F. (2015a). Economic value addition, employment, and enterprise profiles of local authorities in the Free State, South Africa. *Cogent Social Sciences* (2015), 1: 1054610 http://dx.doi.org/10.1080/23311886.2015.1054610.

Toerien, D. F. (2015b). New utilization/conservation dilemmas in the Karoo, South Africa: potential economic, demographic and entrepreneurial consequences. In: G. Ferguson (Ed), *Arid and semi-Arid environments: Biogeodiversity, impacts and environmental challenges*. New York: Nova Science Publishers.

Toerien, D.F. (2017). The enduring and spatial nature of the enterprise richness of South African towns. *South African Journal of Science,* 113(3/4): Art. #2016-0190, 8 pages.

Toerien D. F. (2018a). Power laws, demography and entrepreneurship in selected South African regions. *South African Journal of Science*, 114(5/6): Art. #2017-0255, 8 pages. http://dx.doi.org/10.17159/sajs.2018/20170255.

Toerien, D. F. (2018b). The 'Small Town Paradox' and towns of the Eastern Cape Karoo, South Africa. *Journal of Arid Environments*, 154: 89-98.

Toerien, D. F. (2018c). Productive knowledge, poverty and the entrepreneurial challenges of South African towns. *South African Journal of Science* (in press).

Toerien, D. F. and Seaman, M. T. (2010). The enterprise ecology of towns in the Karoo, South Africa. *South African Journal of Science* 106(5/6): 24–33.

Toerien, D. F. and Seaman, M. T. (2011). Ecology, water and enterprise development in selected rural South African towns. *Water SA,* 37(1): 47–56.

Toerien, D. F. and Seaman, M. T. (2012a). Proportionality in enterprise development of South African towns. *South African Journal of Science*. 108(5/6): 38–47. http://dx.doi.org/10.4102/sajs.v108i5/6.588.

Toerien, D.F. and Seaman, M.T. (2012b). Regional order in the enterprise structures of selected Eastern Cape Karoo towns. *South African*

Geographic Journal 94(2): 1–15. http://dx.doi.org/10.1080/03736245.2012.742782.

Toerien, D. F. and Seaman, M. T. (2012c). Evidence of island effects in South African enterprise ecosystems. In: A. Mahamane (Ed.), *The functioning of ecosystems.* Rijeka: Intech; pp. 229–248. http://dx.doi.org/10.5772/36641.

Toerien, D. F. and Seaman, M. T. (2012d). Paradoxes, the tyranny of structures and enterprise development in South African towns. Presented at: *Strategies to Overcome Poverty and Inequality: Towards Carnegie* 3; Sep 3–7, 2012; Cape Town, South Africa. Accessed at: http://carnegie3.org.za/docs/papers/269_Toerien_Paradoxes,%20the%20tyranny%20of%20structures%20and%20enterprise%20development%20in%20SA%20towns.pdf.

Toerien, D. F. and Seaman, M. T. (2014). Enterprise richness as an important characteristic of South African towns. *South African Journal of Science* 110(11/12). Art. #2014-0018, 9 pages. http://dx.doi.org/10.1590/ sajs.2014/20140018.

Trudon, (2005). *Southern Cape & Karoo Phone Book 2005/6.* Johannesburg: Trudon Pty. Ltd.

Trudon, (2013). *Southern Cape & Karoo Phone Book 2013-2014.* Johannesburg: Trudon Pty. Ltd.

UNESCO, (1996). *Biosphere reserves: The Seville Strategy and the statutory framework of the World Network*. Paris: UNESCO.

Van Waart, S. (2001). Paleise van pluime. Pretoria: Lapa. In *Afrikaans* [Palaces of plumes].

West, G. (2017). *Scale: The universal laws of life and death in organisms, cities and companies.* Kindle edition. London: Weidenfeld & Nicolson.

Wickins, P. L. (1983). Agriculture. In: Coleman, F.L. (Ed.), *Economic history of South Africa.* Pretoria: Haum, Pp. 37–85.

Wikipedia, (2017). Stilbaai Geskiedenis. Accessed at: https://af.wikipedia.org/wiki/Stilbaai#Geskiedenis. In *Afrikaans* [Stilbaai history].

World Commission on Environment and Development, (1987). *Our common future,* Oxford: Oxford University Press.

Youn, H., Bettencourt, L. M. A., Lobo, J., Strumsky, D., Samaniego, H. and West, G. B. (2016). Scaling and universality in urban economic diversification. *Journal of the Royal Society Interface* 13: 20150937. http://dx.doi.org/10.1098/rsif.2015.0937http://www.treasury.gov.za/comm_media/press/2008/Final%20Recommendations%20of%20the%20International%20Panel.pdf).

Contents of Earlier Volumes

Advances in Environmental Research. Volume 66

Advances in Environmental Research. Volume 65

Advances in Environmental Research. Volume 64

Advances in Environmental Research. Volume 63

Advances in Environmental Research. Volume 62

INDEX

A

B

F

G

H

I

J

L

M

N

O

P

R

S

T

U

V

W